STRUCTURAL HEALTH MONITORING
WITH APPLICATION TO
OFFSHORE STRUCTURES

STRUCTURAL HEALTH MONITORING
WITH APPLICATION TO
OFFSHORE STRUCTURES

Srinivasan Chandrasekaran

Indian Institute of Technology Madras, India

World Scientific

EW JERSEY · LONDON · SINGAPORE · BEIJING · SHANGHAI · HONG KONG · TAIPEI · CHENNAI · TOKYO

Published by

World Scientific Publishing Co. Pte. Ltd.

5 Toh Tuck Link, Singapore 596224

USA office: 27 Warren Street, Suite 401-402, Hackensack, NJ 07601

UK office: 57 Shelton Street, Covent Garden, London WC2H 9HE

Library of Congress Cataloging-in-Publication Data

Names: Chandrasekaran, Srinivasan, author.

Title: Structural health monitoring with application to offshore structures /
   by Srinivasan Chandrasekaran, Indian Institute of Technology Madras, India.

Description: New Jersey : World Scientific, [2019] | Includes bibliographical references.

Identifiers: LCCN 2019010512 | ISBN 9789811201080 (hc : alk. paper)

Subjects: LCSH: Structural health monitoring. | Offshore structures--Inspection.

Classification: LCC TA656.6 .C43 2019 | DDC 627/.98--dc23

LC record available at https://lccn.loc.gov/2019010512

British Library Cataloguing-in-Publication Data

A catalogue record for this book is available from the British Library.

For any available supplementary material, please visit
https://www.worldscientific.com/worldscibooks/10.1142/11302#t=suppl

Desk Editor: V. Vishnu Mohan

Typeset by Stallion Press
Email: enquiries@stallionpress.com

Preface

Structural Health Monitoring (SHM) deals with the assessment, evaluation and technical diagnosis of different structural systems of strategic importance. Extensive knowledge of SHM shall lead to a clear understanding of the risk and reliability assessment of structures, which is currently mandatory for structures of strategic importance, such as bridges, offshore structures, etc. A detailed subject matter of this order is scarce in the literature.

The contents of this book showcase both elementary and advanced applications of SHM along with detailed case studies carried out in lab scale. The book addresses the primary requirements of engineering graduates belonging to various disciplines, such as civil, structural, electronics, electrical, instrumentation and communication, mechanical, naval architecture, offshore, marine, and petroleum engineering. In addition, the book will also be a useful reference for engineers practising in the above domain of work and for the faculty teaching this course at the university level. While this course was run as a video course under the auspices of National Program of Technology Enhanced Learning (NPTEL) in India, it invited the attention of a large set of audience from all over the world. The book provides an updated version of the teaching modules and underscores the rich experience of the author in conducting research in this domain.

Chapter 1 of this book presents an overview of structural health monitoring and its application to engineering structures. This chapter deals with the necessity of monitoring the health of structures, with a focus on the various components involved in SHM. In addition to throwing light on the fact that the implementation of the SHM scheme imposes a big challenge in real timescale, details of various factors that influence the implementation process are also discussed. Several components involved in the SHM process are highlighted while specific issues with respect to concrete structures are also discussed in detail. The chapter summarizes the various advantages of SHM along with the long-term and short-term benefits both from economic and safety perspectives.

A detailed perspective of SHM is described in Chapter 2. This chapter deals with the details of different SHM methods and techniques. Damage detection methods that are commonly deployed for civil engineering infrastructure are explained in depth. Damage identification using various methods are also discussed, and a comparison is made between these methods to highlight their applicability issues. Non-destruction evaluation methods with their application issues are also discussed briefly.

Details about various sensor technologies and their application issues are presented in Chapter 3. Various sensor technologies that are commonly deployed in the health monitoring of civil engineering structures are presented. A few sensors that are vital for specific measurements required in offshore platforms are also discussed. The sensor layout and details of the SHM scheme are presented with component-level details.

The application problems of SHM in lab scale are discussed in detail in Chapter 4. This chapter deals with the use of various types of sensors and their limitations. Factors influencing the design of the layout of sensors along with the advantages of wireless sensor networking is discussed. Preliminary design of both wired and wireless sensor networks for the health monitoring of TLP and BLSRP in lab scale are presented. The structural assessment of a bus duct system and the supporting pipe racks for electric transmission lines is also analysed for their seismic safety by imposing a postulated failure. The details of structural assessment, as part of SHM, are presented.

The salient features of the book include the classroom-style presentation of the contents and the experimental studies on SHM both in lab scale and field study. The main objective of this book is to present with detailed explanations and good illustrations of

SHM with applications to civil engineering structures in general and offshore structures in particular. The textbook is presented in a self-explanatory mode, which can be used to teach the course, following the contents outlined chapter wise. Further, this book also elaborates on the experimental investigations carried out in lab scale, in which the SHM implementation is discussed in a stepwise sequence.

I wish to thank the Centre of Continuing Education, IIT Madras, for the administrative support extended during the preparation of the manuscript. I also extend my sincere thanks to team NPTEL for their constant encouragement and support, and finally to all my teachers, students and research scholars, who have been my strength throughout the preparation of this work.

Srinivasan Chandrasekaran

About the Author

Srinivasan Chandrasekaran is currently a Professor in the Department of Ocean Engineering, Indian Institute of Technology Madras, India. He has teaching, research and industrial experience of about 27 years during which he has supervised many sponsored research projects and offshore consultancy assignments both in India and abroad. His active areas of research include dynamic analysis and design of offshore platforms, development of geometric forms of compliant offshore structures for ultra-deep water oil exploration and production, sub-sea engineering, rehabilitation and retrofitting of offshore platforms, structural health monitoring of ocean structures, seismic analysis and design of structures and risk analyses and reliability studies of offshore and petroleum engineering plants. He has also been a Visiting Fellow under the invitation of Ministry of Italian University Research to the University of Naples Federico II, Italy for a period of two years, during which he conducted research on advanced nonlinear modelling and analysis of structures under different environmental loads with experimental verifications. He has about 200 research publications in refereed journals and has also authored 12 textbooks. All the books are quite popular among graduate students of various engineering disciplines and are recommended as reference material for classroom studies and research as well in many universities in India and abroad. He also delivered 12 numbers of web-based courses through National Program on

Technology Enhanced Learning (NPTEL), a government of India initiative to disseminate quality education through open domain. These web-based video courses are very popular amongst academia and practising engineers. He is a member of many national and international professional bodies and delivered many invited lectures and keynote address in the international conferences, workshops and seminars in India and abroad.

Contents

List of Figures

List of Tables

Chapter 1

Structural Health Monitoring: An Overview

This chapter deals with the necessity of health monitoring of structures, focussing on the various components involved in structural health monitoring (SHM). While implementation of the SHM scheme imposes a big challenge in real timescale, the details of various factors that influence the implementation process are also discussed. Several components involved in the SHM process are highlighted, while specific issues with respect to concrete structures are discussed in detail. This chapter also summarises several advantages of SHM along with the long-term and short-term benefits both from economic and safety perspectives.

1.1. Introduction

Civil engineering structures in general and offshore structures in particular attract huge capital investments apart from demanding a longer period of time for their construction, installation and commissioning. Continuous monitoring and proper maintenance of structures are vital requirements of the modern society. It is no longer optional for the owner to upkeep public buildings in terms of its functional maintenance, and legal mandates have also been imposed on the safety and security of public buildings; in this aspect, health monitoring of infrastructure is also supported and funded by government agencies. While advanced methods of analysis and design of structural systems, which are in compliance with the code guidelines, confirm least probability of design failure, construction difficulties pose various challenges, resulting in deviations to the design solutions causing uncertainties. In addition, ageing of structures poses additional challenges in terms of strength degradation and non-availability on demand.

In addition to helping engineers recognise poor structural conditions and other safety issues, advancements in SHM also help professionals determine potential risks of buildings caused due to ageing and other environmental factors. SHM is particularly useful in preventing water and flood damage caused by failed dams, dykes, pipelines, and other similar structures. SHM essentially examines the state of the present condition of structural systems to assess their functional fitness and performance levels. If the condition assessment envisages a poor performance level in comparison to that of the desired ones, repair processes are immediately initiated. It is therefore emphasised here that structural repair is invoked only when the structural system undergoes significant damage; but such practices are not acceptable in the case of structures of strategic importance, such as nuclear power plants, large reservoir dams, offshore and coastal structures. Hence, in order to plan for a schedule of preventive maintenance, both in terms of shutdown time of the facility and economic planning, SHM is used as one of the essential tools in reliability engineering. Further, SHM also enables the reduction of long- and short-term costs with respect to maintenance and repair of public building; this reaps the economic benefit of the construction industry to a great extent.

Large, complex and costly engineering structures are constructed to last for a longer life span. While the usual practice is to design them for maintainable conditions, their design life is usually extended even under non-maintainable conditions. The general scope of structural health monitoring (SHM) includes structural assessment, monitoring and control, which can be abbreviated as SAMCO; all the three components are vital in SHM. Structural assessment involves the assessment of actual conditions and load-carrying capacity of the structural systems. Diagnosis is a vital part of SHM, which involves integration of various sensors used for measurements, computational power and processing ability within the SHM system itself for effective outcome. Structural monitoring deals with the supervision of structures on a continuous basis using sensors or electronic gadgets. They are carried out in order to maintain the functional utility of the structure; to be very precise, it is done to ensure the availability of the system on demand. Thus, it includes both periodic and preventive maintenance. Structural control deals with the control of the dynamic response behaviour of structures under various environmental loads. This involves the establishment of control mechanisms so that responses are under the preferred limits even on unforeseen increase of load magnitudes. The above three vital components can be prioritised as follows:

- Assessment involves the preparation of existing conditions of structure in terms of its geometric fitness and load capacity and therefore deals with the examination of those conditions of the structure that one can take advantage of beyond its design life.
- Monitoring is more or less related to maintenance, ensuring the availability of the structural system on demand. Monitoring of structures does not necessarily mean knowing the status of structures in real time. Since all structures are designed with a margin of safety, which can be used to exploit its design life to at least a marginal extent, tolerance can be allowed in delaying maintenance. Monitoring, therefore, enables effective planning of maintenance.
- Control aims to reduce or mitigate undesirable (modes of) response even under unforeseen increase in magnitude of loads. This is very vital in the case of offshore compliant structures as their load resistance is often derived from FORM resistance and not from their strength. Excessive displacements in global degrees of freedom, as a rigid body motion, may become undesirable as this may even challenge their hydrodynamic stability. Hence, control of both rotational displacements in particular and translational displacements in general is very important for ensuring structural safety and safe operability.

Priority depends upon two factors: (i) type of the structure and (ii) economic considerations, which drive the whole concept of application of SHM in structural engineering. In view of a normal type of structure under given budgetary normal considerations, assessment of the condition of the structural system is very important. In order to develop an effective design, one can even prescribe control mechanisms; but, prior to that, maintaining the utility value of the structure to ensure its availability is also equally important.

SHM actually deals with the development and implementation of methods and techniques, which are useful for ensuring the availability of the structural system to perform its intended function on demand. Thus, the main objective is neither exercising a control algorithm nor the assessment of load-carrying capacity but ensuring the functional utility value even under critical environmental conditions. Continuous monitoring (or even periodic monitoring to a larger extent) ensures preventive maintenance and helps policy and planning guidelines to ensure the functional value of strategically important structures. For example, a periodic and preventive maintenance of a highway bridge or gas pipeline shall ensure uninterrupted service and minimum downtime, even in the case of any critical repair. Civil engineering professionals agree to the fact that maintenance of infrastructure facilities is vital in order to elevate the standard of the structural systems in terms of their serviceability, appearance and safety. As certain clauses of structures, such as industrial structures, highway and railway bridges, nuclear power plants, offshore structures, naval structures, etc. are vital for the economic growth of the country, ensuring their availability on demand is necessary for the safety and security of public life. They also influence the economic growth of the nation in the international market. Society essentially depends on these structures for various reasons, such as economic, environmental, life-quality updates, safety and employment perspectives. Most of such structures also reach critical age, which can result in strength degradation, degraded quality of appearance, decreased load-carrying capacity and reduction in the overall dependency. In order to ensure and continue a comfortable dependency on these structures, both periodic and preventive maintenance are important. There are three ways by which maintenance can be attempted:

(i) periodic maintenance; (ii) preventive maintenance; and (iii) critical maintenance (maintenance on demand).

Critical maintenance is more alarming and dangerous, where a structure is maintained only after its critical age. In this case, recovery of strength of the structure is very difficult. For example, let us consider offshore structures used for oil and gas exploration and production. An offshore platform working continuously on oil and gas production results in an outcome or commercial benefit, which could be revenue, employment, constant research and development for further exploration and production. If, due to unavoidable reasons, structure needs to be shut down for maintenance, the shutdown period (known as downtime) will primarily lead to loss of revenue, which is not preferred. On the other hand, if a total shutdown could be avoided by way of preventive maintenance, it can lead to several advantages; economic benefit is the foremost one. When the structures reach critical age, there can be strength degradation due to material corrosion in sea. Thus, these structures will not be able to alleviate the encountered lateral loads successfully. It may also result in structural failure, which can cause disaster. One cannot afford to lose such novel, unique and high-investment structures. Preventive maintenance can avoid such catastrophic failures. To carry out preventive maintenance, one must assess the present condition, monitor the condition continuously and then plan the repair procedure even before the structure actually needs it. Instead of doing a periodic maintenance, strategic structures can demand a preventive maintenance. Preventive maintenance is possible only through detailed assessment, monitoring and planning of repair, which are all encompassed under the framework of structural health monitoring. Therefore, one of the most important deliverables and main outcomes of SHM is the avoidance of a premature failure or a breakdown of the facility.

Let us consider another example: a naval dockyard, which is essentially an open channel to house large vessels for their periodic maintenance. Periodic maintenance could be partial or complete weld upgrade, painting, treatment for biofouling, upgrade fault correction for electromechanical systems, etc. Navy operates on various kinds of strategic vessels like submarines, which need to be inspected for periodic maintenance or emergency fault correction; such operations are usually carried out in a dockyard. Dockyards are very few in number and quite expensive. If a dockyard is undergoing periodic maintenance, which demands the shutdown of operation during an emergency requirement of docking a naval vessel, functional assurance of the essential service becomes unguaranteed. Hence, the shutdown time caused by a periodic maintenance schedule on a dockyard deprives the basic utility value of the system itself; this can be avoided if the dockyard undergoes preventive maintenance. Maintenance should be carried out in a preplanned and preventive manner, so that the dockyard always remains functional even during critical environmental conditions. However, utility value can be slightly decreased in terms of its operational capacity, but a complete shutdown of the dockyard can be avoided. Hence, preventive maintenance of essential services is far more advantageous in comparison to periodic maintenance. SHM ensures constant maintenance costs and high degree of reliability of the service instead of high maintenance cost and low degree of reliability.

1.2. SHM Analogy

When a human being falls sick, he is physically examined by a medical doctor. Similarly, inspection demands a complete analysis of the structural condition, which can be done by structural monitoring. Further, in the case of a human being, health can be monitored on a continuous basis through several means under any situation as per the advice of the doctor. In a similar manner, structural health monitored using sensors records the response of the structure through a typical time-history response plot. For example, accelerogram is the graphical output of seismograph. In the case of human health monitoring, when one undergoes electrocardiogram, plots are available, which indicates the health condition of the human being. So far, the person is referred to as human and not patient because he is not sick. This process of health monitoring is called "diagnosis" in medical terms and "monitoring" in SHM.

Monitoring the output could result in the assessment of the condition of the structure and enable one to reach a conclusion about the present condition of the structure; this is called as "assessment". Similar to the post-diagnosis report, which clarifies the state of health of the human being and helps seek expert medical advice, SHM helps one to recommend certain control algorithms for reducing (or completely mitigating) the undesired responses of the structure; this is termed as "control". Based on the diagnosis and assessment, the doctor may recommend a surgery, which attempts to mitigate the problem. In short, both human health monitoring and structural health monitoring result in ensuring overall safety and satisfactory functionality of the system (or human) as the case may be; hence, both processes follow the same analogy.

1.3. Necessity for SHM

The overall objective of SHM is to ensure a satisfactory performance of the structural system in its present condition. Infrastructure investment is always not only towards new construction but also includes its maintenance. There can be a slack-down time in infrastructure growth under economic recession during which investment towards maintenance may become important. During such periods, major investment can be directed towards the maintenance of old existing structures. Structures that have reached critical age (may be 30−40 years of service life) demand higher order of maintenance to upkeep their functionality. The deterioration of the structure's condition depends on various factors: (i) type of material; (ii) nature and type of loading; (iii) environmental conditions; and (iv) degree of maintainability. However, in general practice, structures with 30−40 years of service life fall under the category of critical ageing, which requires adherence to a periodic inspection and a systematic maintenance schedule. The state of health of the structure and the damage are assessed by a few closely related disciplines, which include structural health monitoring (SHM), non-destructive evaluation or testing (NDE or NDT), condition monitoring (CM), health and usage monitoring system (HUMS) and damage prognosis (DP).

SHM involves the implementation of damage detection strategy for structures of high importance. Condition monitoring is similar to that of SHM but commonly used in mechanical and power generation systems. NDT is more a traditional technique, which includes visual inspection, liquid penetration, ultrasonic, X-ray, radiography, eddy current methods, magnetic field methods, etc.

Under the expected rise in the repair and retrofit segment in the near future, construction industries should be prepared with methods, strategies and technological skills to carry out repair. It is possible only when SHM is in existence. SHM is essentially a process of developing and implementing damage identification strategy, which involves the identification of a statistical pattern affecting the present and futuristic conditions of the structure. It is therefore necessary to train the technological manpower, who can take care of immediate repair and retrofit procedures for the structures, which demand this kind of attention. SHM deals with the preparedness of carrying out repair and retrofit of structures that demand special attention. One of the major and successful outcomes of the healthy practice of SHM is disaster prevention. One can completely mitigate disasters caused by natural events if the structures are monitored on a continuous basis. It is also important that they are maintained to upkeep their functional value with respect to their present age and working conditions.

One of the main objectives of SHM is to fulfil the necessity of a disaster prevention mechanism. Natural disasters such as earthquakes, Tsunamis and cyclones have demonstrated the vulnerability of buildings, coastal structures, nuclear reactors and other structures of strategic importance under the unexpected environmental forces. Thus, natural disasters not only lead to loss of life but also challenge the economic sustainability of the nation. Hence, the first necessity is the preparedness for such natural calamities, which is followed by ensuring economic sustainability and knowledge update. For example, recent earthquakes taught interesting lessons of various failure scenarios, so that appropriate design procedures and ductile detailing are enforced through the design codes. This is possible only when there is a constant update about the loss of strength of structural systems under unexpected forces like earthquakes. This is termed as monitoring, which is actually a vital part of SHM.

1.4. Scientific Justification

In the overall domain of SHM, the following steps may be of fundamental interest:

- Identify structures that need monitoring.
- Acquire information about probable degradation of materials and risks involved in the structural system from the designers.
- Establish expected responses of the system to these probable degradations.
- Design a proactive SHM system, which can detect such conditions through a carefully integrated sensor network.
- Install and calibrate the SHM network.
- Acquire, analyse and manage data.
- Schedule a proper emergency response plan in the case of any emergency that arises from non-functional responses of critical infrastructures.

Damage detection during the early stages requires continuous monitoring. SHM encompasses a process of implementing damage identification strategy to the offshore, civil, mechanical or aerospace structures (Farrar and Worden, 2007a–c). It can be either global or local assessment. Global assessment is when the structure is assessed as a whole. Based on the response of the whole structure, the damage and remaining life of the structure are assessed. On the other hand, in local assessment, each member is examined independently and maintenance approach is carried out. SHM, therefore, evolves time-based maintenance approach, which is an important outcome of condition assessment. In time-based maintenance approach, the maintenance process is proposed to be carried out at specific intervals of time, irrespective of the current state of the system. If a member fails in between, then it has to be replaced. Such interventions shall increase the downtime of the system if the repair operation requires sufficiently longer lead time. This type of process includes tradeoff between the cost and the risk, which arises as a consequence of the damage. This shall also account for unexpected variations in load and environmental conditions that prevail on the structure. Most of the private and a few government industries show interest in detecting damages in their developed products and manufacturing infrastructure, so that damages can be detected at the earlier stages of process/production. Such detection requires intensive use of SHM, which is motivated by potential life-safety and economic impact of this technology.

SHM has been successfully deployed in the oil industry, large dams and highway projects, whose installations have been noted with careful attention and appreciation to the research efforts. However, a few of the heritage structures in Italy (see, e.g., Lady of Shrine, Siracusa) are also kept in the loop of SHM network due to their cultural importance. According to the International Society of Structural Health Monitoring and Intelligent Infrastructure (ISHMII), a lot of bridges and structures are monitored using various sensors for any damages; application is quite common and popular in European countries due to the fact that Europeans have a culture of retaining old structures in a good condition. Some Asian countries have also seen a rapid growth in health monitoring culture: Japan, Taiwan, China and Singapore, where monitored structures are quite healthy and safe from the functional perspective.

The use of SHM in assessing occupational safety of buildings under natural calamities like earthquakes is quite vital. For example, as such, there exists no quantifiable method to determine occupational fitness of a residential building after a significant earthquake. SHM can be seen as one of the scientific tools to minimise uncertainty associated with post-earthquake damage assessments. Similarly, a dockyard, after being serviced for a vital repair, needs to be assessed and continuously monitored for the successful recovery of strength and functionality of the dockyard. If SHM methods can assure prompt reoccupation of industrial buildings and critical infrastructure like railway bridges, highway bridges, etc. after significant earthquakes or floods, this can help mitigate economic losses associated with such natural calamities. While it is a fact to understand that most of the current structural and mechanical systems have crossed their design life on a time-based model, buildings of heritage value are overdone for sure. Hence, SHM technology will allow the current time-based maintenance policies to evolve into potentially cost-effective condition-based maintenance policies. But, in condition-based maintenance approach, actual state of the structure is taken into consideration for conducting the maintenance process. Such a process, when implemented, will reduce the downtime. In addition, it increases productivity, reduces life-cycle cost and increases safe operability (Liu and Nayak, 2012). SHM enables preventive maintenance operations when there is a likelihood of response exceeding the threshold value. SHM activities involve a five-level classification as follows (Rytter, 1993), (Keith *et al.*, 2003):

- First level is the assessment of the response to determine whether the structure is damaged or not.
- If damaged, it further tries to identify the localisation of the damage.
- Based on the data, it will quantify the amount of damage.
- It shall also predict future progress of the damage and remaining service life.
- It further recommends appropriate remedial and repair measures to restore both strength and functionality of the structure.

Modern world depends on complex and exhaustive systems of infrastructure. Many structures were constructed during the economic progress in the recent past. All these structures are now aged. According to statistics, in many advanced countries, more than 40% of the bridges are critically aged (found to be older than their design life). In general, the public funds available are too less for the replacement of the structure. It can only enable partial repair of the structures. In such cases, one needs to know the justification for the partial repair, which can be established only through SHM. Using effective approaches, a regular periodic maintenance can also be planned effectively. Thus, effective planning of maintenance also requires a continuous monitoring of the condition, which is an essential outcome of SHM. Therefore, SHM is completely a scientific approach involving the capability to understand the importance of successful maintenance of civil infrastructure. Further, SHM also involves the use of various automated tools and systems, which are used to improve the inspection procedures and techniques of repair. Guniting is one of the methods of repair, which can give surface treatment for material degradation against corrosion. The scientific approach of SHM can improve safety standards of public life. It can reduce risks and enable us to discover new methods of reducing cost of repair and rehabilitation.

Vibration-based damage detection techniques are commonly used in damage diagnosis as they are one of the most efficient methods. There are various physical parameters; microelectromechanical system (MEMS) has its own advantages and is applied in various fields of application, such as biomedical, automotive, construction and consumer sectors. Vibration-based damage detection methods are further classified into traditional and modern approaches. The traditional method is based on the principle that change in mass and stiffness will be reflected in the measurements of natural frequency and mode shapes of the structure. When the measured data of natural frequency or mode shape are different from that of the normal, it indicates the initiation of damage. The modern method involves the online measurement of structural response to detect damage with the help of signal processing techniques, artificial intelligence and neural networks (see Table 1.1). Dynamic response of the structure under different loads is measured online while SHM

Table 1.1. Sensors for vibration monitoring.

Physical parameter	Sensing principle	Technology
Acceleration	Inductive sensors	Conventional
Velocity	Capacitive sensors	MEMS
Displacement	Piezoelectric sensors	
Magnetic field	—	Giant magneto-resistance
Magnetic resistivity		
Optical property	Photoelectric sensor	Fibre Bragg grating
	Optical fibre sensor	Fabry–Perot interferometer
		Intensity sensor
Acoustics	—	Ultrasonic probes

indicates change of structural parameters, thereby detecting damage in the structure.

1.5. Major Advantages

SHM possesses several challenges, the foremost challenge being optimal definition of sensors in terms of its choice, type, layout and number of sensors to be deployed. The next challenge comes from the communication systems whether these sensors will be wired or wireless, whether the communication will be through R/F or other modes of transformation. There are some salient advantages of deploying the SHM scheme. These advantages are common to a variety of structures, such as civil engineering structures, mechanical systems, offshore structures, naval systems, aviation systems and nuclear reactors. The advantages are listed as follows:

(1) The major advantage, which is mostly welcomed by the engineering fraternity, is that the SHM scheme enables one to update the integrity of the structure. This is true if monitoring is carried out on a continuous basis.
(2) The utility value or functional value of the structure is enhanced. It means the structure is put to its optimal use.
(3) It minimises the downtime. A preventive maintenance can be preplanned ahead based on the monitoring and assessment of the structure. This is very helpful in naval defence systems.
(4) Public safety is enhanced. For example, if SHM is deployed on a bridge and monitored continuously, then its functional ability is predicted or assessed to a higher accuracy to avoid any catastrophic failure.
(5) There is a significant improvement in maintenance organisation of public structures. One can avoid unnecessary maintenance schedule. Critical elements which require immediate attention are not ignored under SHM deployment. It enables one to carry out periodic maintenance with performance-based focus. This a very recent trend which enables a lot of cost-saving and enables an engineering efficiency in preplanning maintenance in terms of structures of very high strategic importance.

(6) It reduces investment on maintenance labour. Inspection labour is expensive, it is also very special and technical, and it is time consuming. All these can be avoided by the regular maintenance schedule. SHM reduces human involvement towards inspection, planning and decision making on maintenance schedule. Maintenance is actually planned, scheduled based on the structural condition automatically. Unsatisfactory maintenance has many critical disadvantages. The consequences that arise from unsatisfactory or improper maintenance cause further disaster, e.g., the accident of Aloha airlines and the collapse of the Mianus river bridge. Efficient use of funds towards maintenance is reduced and time or schedule of maintenance can result in downtime of the facility despite its critical need, e.g., dockyards.

One of the major advantages of using SHM is that it includes the reduction of cost related to inspection and mitigation of impact of structural disasters caused by nature. Further, it reduces the need for immediate repairs and thereby improves public safety. From an overall perspective, it improves cost efficiency of public funding in a more reasonable manner, as discussed in the following sections.

1.5.1. *Increased safety*

SHM practices ensure improvement in public safety. SHM also ensures effective utilisation of public funding towards the maintenance of civil infrastructure of any nation. SHM is also advantageous in the replacement of water supply pipelines which had severe metallic corrosion. Preventive maintenance in this case enhances the quality of public life. It ensures the use of new tools and technologies to carry out and maintain serviceability of structures and also helps us to even declare them as safe or unsafe. In the case of ageing structures, SHM is advantageous because the health of the structure can be monitored using sensors, data collection and analysis to initiate a preventive maintenance. Further, continuous monitoring and analysis of the recorded data helps to update design procedures by resolving any flaws in the design. It also serves as a knowledge update on the design of structures.

1.5.2. *Detection of early risk*

SHM tools can be deployed to detect a poor structure or its condition, and therefore its usage can be limited. This enhances public safety. Secondly, SHM can be seen as a highly useful tool in preventing water and flood damage caused by failure of big reservoirs. In such cases, built-in sensors will be useful to monitor the change in water level which can be used to detect minor leaks and major failures as well. SHM can also be used as a new design tool in the case of design of foundations for bridges, pavements, etc. To a reasonable extent, ground movement can be monitored, which can help us in predicting earthquakes. This will improve the preparedness of the structures during the forthcoming earthquakes.

1.5.3. *Longer life span*

Both preventive and periodic maintenance enhance the service life of the civil structural systems. Continuous monitoring improves the plan for preventive and repair procedures. Most importantly, it accounts for human errors if made. SHM can also improve the existing design methods by eliminating the flaws in the design procedure. This enhances immediate safety in public buildings.

1.5.4. *Cost efficiency*

SHM can be helpful in the effective utilisation of public funding towards maintenance. It can essentially avoid unwanted maintenance of structures with good health, i.e., an unnecessary periodic maintenance of a system already in good health can be avoided. It avoids shutdown of operations as explained earlier, which can enhance the economic efficiency of the system by enhancing the return on investment (ROI) in the case of oil and gas industry.

1.6. Components of SHM

SHM deals with continuous monitoring, assessment and then control algorithm to be in place for establishing a satisfactory performance level of a given structural system or any infrastructure. Considering this as one of the important objectives of SHM, let us look at the components involved in SHM. The components of the SHM process consists of different stages: (i) operational evaluation; (ii) data acquisition (DAQ), fusion and cleansing; (iii) feature extraction and information condensation; and (iv) statistics-based model development for feature identification. Operational evaluation is focussed on a few vital points: (i) damage definition, i.e., under what condition, it is said to be damaged; (ii) economic issues; (iii) data management; and (iv) environmental or operational constraints, if any. Data acquisition, fusion and cleansing stages deal with the sensing and data acquisition issues. It is focused on determining the method of measurements, such as strain, displacement, acceleration, temperature, wind speed, sensor placement and other related issues. It also deals with various possible excitation methods that can be deployed, such as ambient excitation, forced vibration or local excitation, and the type of data transmission, such as wired or wireless. Various parameters and methods are used in feature extraction and information condensation, including resonant frequencies, frequency-response function, mode shapes and mode-shape curvatures, modal strain energy, dynamic flexibility, damping, Ritz vectors, extracting nonlinear features, empirical mode composition, wave propagation, Hilbert transform, evaluation of auto-correlation function and other related features. Statistical model development involves two methods: supervised learning and unsupervised learning. Supervised learning includes response surface analysis, Fisher's discriminant, neural networks, genetic algorithms and support vector machines. Unsupervised learning includes control chart analysis, outlier detection and hypothesis testing.

In the case of health monitoring of a bridge, as shown in Figure 1.1, there will be varieties of sensors placed on the deck of the bridge. The sensors are the first-level components of the health monitoring system. In parallel, let us also look at the health monitoring of an offshore structure with multi-tier deck with drilling derrick, moon pool, living quarters, flare boom, helipad, etc. The offshore platform is supported on a template structure founded to the sea bed, as shown in Figure 1.1. There may be different varieties of sensors placed on deck and living quarters at various levels above

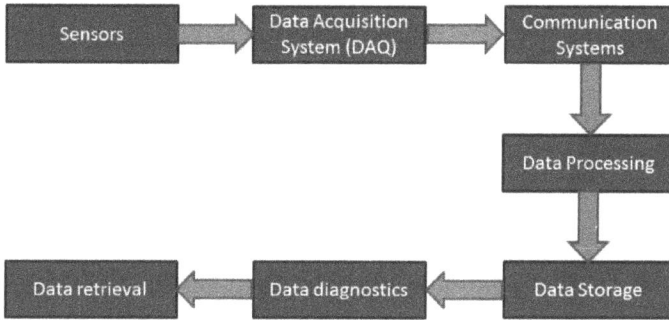

Figure 1.1. Components of SHM.

the water level. All these sensors need to be connected to a DAQ system, which is the second component in SHM. From DAQ using communication systems, they will be transferred for data processing. The communication system is the third component in SHM, which will collect the data for post-processing. The data processing is the fourth-level component in SHM. Once the data are post-processed, the data need to be stored; data storage is the fifth stage in SHM. It is also known as data repository. Stored data will be taken to data diagnosis and data retrieval. The layout in Figure 1.1 includes various components, which are involved in a complete health monitoring system used for infrastructure engineering starting from sensors to data retrieval. From the complete layout of SHM, one can divide them componentwise.

The vital components of SHM are as follows:

(1) **Sensors:** There are different types of sensors based on layout of topography, scalability, etc.
(2) **Data acquisition:** It depends upon the type of DAQ used, whether it is going to handle wireless or wired sensors, etc.
(3) **Communication system:** It can be done using either R/F frequency or Intranet.
(4) **Data processing:** It deals with the statistical analysis of the collected data.
(5) **Data storage:** It also requires data diagnosis and data retrieval.

Application complexities of various components with respect to different industries, such as aviation industry, civil infrastructure industry, mechanical industry, oil and gas petroleum industry, will be discussed later.

1.7.   Sensors Used in Health Monitoring

(a) **Fibre Bragg diffraction grating sensors:** These are embedded in structures, which are laser marked with optical interference parameter. They measure the local strain caused by the deformation, which results in sensor measurement. These sensors will transmit a different wavelength, based on which the measured deformation can be detected.

(b) **Acoustic emission sensors:** These work on the basis of acoustic signals, which are generated by the presence of cracks or local faults. These sensors are useful to measure delamination of fibres or breakage.

(c) **Smart sensors or sensor coatings:** These are paints or coatings, which are applied on the surface. They remain integrated with the piezoelectric or ferroelectric elements to measure the strain variation. Sometimes, carbon nanotubes are also used to detect such variations. A detailed spectroscopic analysis is required to process the strain variation caused by the damages in the local scale.

(d) **Microwave sensors:** These are actually useful to indicate moisture ingression when embedded in structures. They are useful and efficient in composite structures.

(e) **Imaging ultrasonic sensors:** These contain an ultrasonic wave transducer, which generates a signal that passes through the material. Change in reflection indicates the flaws, presence of cracks or any other local damage.

1.8.   General Challenges in SHM

Following are a few challenging scenarios that occur very often while implementing SHM:

- The foremost challenge in the SHM industry is the development and demonstration of the health monitoring technology, which can be useful to maintain the structural integrity with improved reliability and durability. There are many techniques by which health monitoring can be carried out and is being practiced all over the world. Undoubtedly, most of them are successful as well. However, at one point, it is agreed that developing a technology, which suits a specific application problem, is one of the most important and major challenges in the SHM scheme. Unlike conventional non-destructive techniques, a single technology of health monitoring cannot be suitable for all applications, which makes it more challenging. It depends on various factors such as material, component geometry and identifiable damage scenarios of a given structural system.
- Further more challenging is the fact that the outcome of monitoring scheme should be reliable because sometimes it may trigger an unwanted maintenance which is expensive. It may also sometimes create spurious warnings which should be avoided. Such situations generally degrade the confidence level on the strength of the existing structures.
- Next issue is the optimisation of structural design on the basis of data acquired or monitored through SHM. It is important that this data which is acquired through SHM should be fairly accurate and robust.
- The next challenge could be the major concern regarding the cost of the whole scheme.
- Furthermore, an important factor is related to the owners of the structural system. Even in the case of government undertaking schemes, the use of SHM needs to be scientifically established as it invokes public funding on a major investment. Therefore, it should produce a reasonably advantageous outcome from economic and public safety perspectives. It should be producing results that are debatable and comparable to the regular maintenance approaches.
- The next major challenge is actually the damage detection itself. This is related to the location of damage, origin, scalability, prospective growth and consequences.

- The other major challenge is reliability and robustness of the sensors, their lifespan, and adaptability to working environment and successful suitability to sensor network.

Some of the additional challenges in the SHM scheme include damage identification in civil and mechanical structures. For a given structural system, locating the damage itself requires a lot of experience and database comparison in order to identify the parameters causing such damage initiation.

1.8.1. *Comparison of structures with and without SHM in terms of reliability*

Figure 1.2 shows the comparison between two sets of operations, where one set of operation is with SHM deployment and other set without SHM. The quality of the structure in terms of its functional value is given by the reliability level and the maintenance cost. If there is a system which is without SHM, reliability of the system will be very high initially and then it decreases with the decrease

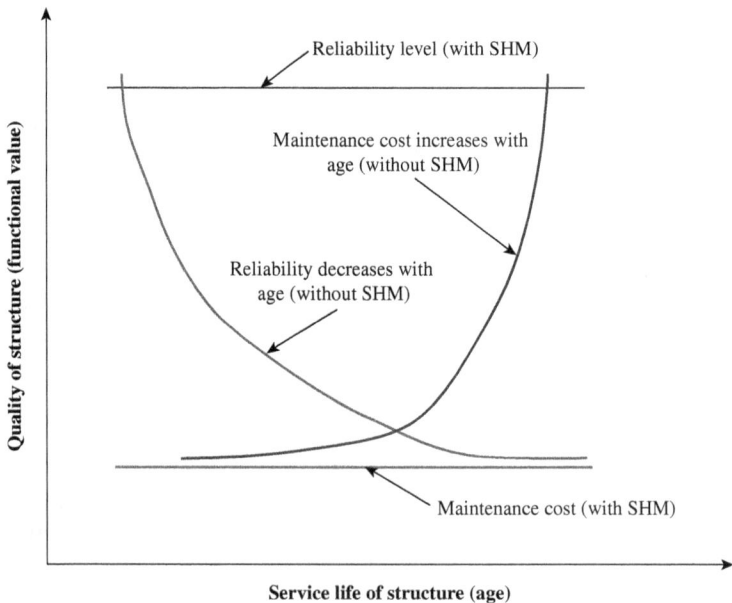

Figure 1.2. Comparison of structures with and without SHM.

in quality as the service life increases. This indicates that there is a reduction in reliability with age for structures without SHM. But the maintenance cost will also be increasing with age for structures without SHM. On the other hand, for the structural systems with SHM, reliability will keep on increasing with the age of the structure. Here, the system will have more or less constant maintenance cost and standard reliability level. So, deployment of SHM in terms of quality of the structural system and enhancement in service life can be achieved with more or less standard maintenance cost or lower maintenance cost, provided SHM is deployed using proper methods. Thus, structures with SHM deployment show more or less a constant maintenance cost, even with the increase in the service life of structure or ageing. It also shows a constant reliability level, indicating assurance of acceptable quality of the structure in terms of its functional value. Of course, without SHM, there is an increase in maintenance cost and decrease in reliability with ageing.

1.9. SHM Methods

Classification of SHM essentially depends upon the techniques used for damage detection. There are four levels of damage identification: (i) Level I, which is focused on the determination of damage in the structure; (ii) Level II, which highlights the determination of geometric location of the damage; (iii) Level III, which quantifies the severity of damage; and (iv) Level IV, which predicts the (remaining) service life of the structure after occurrence of damage. Based on the levels of damage, SHM can be executed in four stages: (i) operational evaluation; (ii) data acquisition; (iii) extraction of information and data processing; and (iv) development of appropriate statistical model for feature discrimination. We shall discuss these in detail in the following sections.

1.9.1. *Operational evaluation*

This stage consists of various factors, such as economic consideration, life-safety issues, definition of damage, details about environmental constraints, operational constraints, data collection and management.

1.9.2. *Data acquisition*

This stage depends upon the following aspects:

- **Excitation methods:** The choice of data acquisition system also depends on the type of excitation which includes forced excitation (number of channels, frequency and bandwidth depend on the type of excitation), ambient excitation (in this case, frequency can be low), and local excitation (forces will be of lower level) and global excitation (larger forces).
- **Data transmission:** Wired or wireless.
- **Sensing the structural responses:** This includes the types of responses measured from the structure, such as strain, displacement, acceleration, temperature variation, wind force and wave force.
- **MEMS technology for sensing.**
- **Fibre optic sensors.**
- **Sensor layout and location of sensors**, which includes scalability and power management.

1.9.3. *Feature extraction and condensation of information*

This is also referred to as data management and processing. There are various parameters and methods, which are used to extract the vital information which are required to assess the present health of the structure. A few of them are listed as follows:

- resonant frequency band;
- frequency response function;
- mode shape;
- mode shape curvature;
- modal strain energy;
- dynamic flexibility;
- damping introduced in the structure due to defect;
- anti-resonance characteristics;
- Ritz vector;
- canonical variant analysis;

- nonlinear features;
- time frequency analysis;
- empirical mode decomposition to know the higher order contributions;
- Hilbert transform;
- wave propagation;
- auto-correlation function.

1.9.4. *Development of statistical model*

This helps in identifying the vital parameters in assessing the structural health. This can be subdivided into two types:

- learning under supervision;
- unmonitored learning.

Learning under supervision deals with response surface analysis, fissures discriminate, neural networks and genetic algorithms, whereas learning under unmanned conditions deals with control chart analysis which is more or less automatic, outlier detection which will filter out the outlayered values in the recorded excitations, neural networks which also has the capability to train the system without supervision and hypothesis testing. Out of all the four components of the SHM process, the most difficult task is the choice of statistical model which helps us to extract the information related to assessment and monitoring.

1.10. SHM: State-Of-The-Art Application

Every system, whether it is mechanical, electrical or structural, has minor defects; this may be due to manufacturing constraints, poor quality control of material used for manufacturing or construction, improper design, effects caused by unforeseen environmental loads, etc. As long as these defects are minor and does not interfere with the functionality of the system, it may be acceptable; but they can continue to grow and result in a sudden failure, leading to catastrophic consequences. Such disasters can be predicted and avoided by deploying SHM; this essentially reduces to identifying defects (or damages) in the (new or old) system. While there are many methods available for damage detection, the SHM process depends largely on the technique used for damage detection. Therefore, identification and location of damage are very important, but no single method of SHM can address these problems, which can be commonly applied to all types of structures. This means that different techniques of SHM are practised and they all have damage-related dependencies.

One of the important factors in the SHM identification is sensitivity. Highly sensitive techniques may show false-positive positions of damage location; low sensitive techniques may show false-negative positions. Therefore, sensitivity of the sensors deployed is again problem-specific. Further, lifetime prediction of service life based on damage modelling is actually very difficult. Most of the techniques which use the service life prediction based on damage modelling are based on reduction and rigidity of the member, but reduction and rigidity must be related to strength. Otherwise, they are not useful for reliability estimates, which are an essential outcome of SHM evaluation.

1.11. Key Issues in Choosing SHM

There is a wide variety of non-destructive testing (NDT) methods and visual inspection techniques that are useful replacements of the SHM system. This is due to the main fact that sensors, which are permanently embedded in the SHM systems, make the process expensive. But, there exists a major advantage of using SHM in place of NDT methods. Measurements observed by the sensors in the SHM process are interpreted by software, which is responsible for processing and managing sensor information. One of the major differences between the traditional NDT and SHM systems is the integrated approach and autonomous inspection that are adopted in the SHM process, which makes the structures more intelligent. SHM is actually not a commodity to purchase but need to be designed and developed. SHM is problem-specific and it cannot be a generic system. High engineering cost and lack of resource availability leave no better choice for the designer, except to choose one of the existing health monitoring schemes or systems.

Unfortunately, most of the SHM systems rely on the point sensors. Point sensors obtain data at one point to monitor. They have a few limitations. Primarily, it is with respect to their insight and not with respect to their accuracy and reliability as these sensors are perfect, scientifically advanced and highly ultra-modern. When there is an event that occurs between the critical points where point sensors are installed, major information about the structural health will be lost. Secondly, it is with respect to the data normalisation. Data normalisation is the process of separating the data occurring from different changes in the behaviour of the structural system. This is essential as the sensor output contains combined information, which is complex to separate. They will include the damage caused by the environment, structures or material degradation, making it complex through combined representation. Therefore, non-continuous monitoring will not be helpful to normalise the data. The solution could be the usage of fibre optic sensors (FOS), whose application is well established by SENSURON in Europe. FOS can be fully automated to detect the local damage through continuous monitoring. Therefore, the system relies less on the interpolation of data. Since the data are continuous, it is also easy to attribute the changes that arise from various conditions, such as environmental factors, material degradation, etc. The use of FOS has long-term benefits and possesses a high degree of convenience to use.

As discussed in the International SHM workshop, SHM development can be divided into the following sequential steps: (i) detection; (ii) identification; (iii) quantification; and (iv) decision (Fu-Kuo Chang, Stanford University). While it is observed that detection is the lowest level of maturity that the SHM system should (at least) posses, a good quantification should lead to a correct decision and efficient solution. One of the most expected innovations in SHM could be bio-inspired sensor networks, which include a large number of small sensors, switches, processors with the accompanying software. If implemented in a massive scale, this will result in a higher reliability apart from being highly economical due to the large production volume and the capability of being installed in a large variety of structures; C-MOS and MEMS manufacturing techniques are lead-liners towards this modernisation.

1.12.   Uncertainties in the SHM Process

Old existing structures do not show up any defect or deficiencies until they experience a disaster. However, it would be too late by then as the damage would have already occurred in terms of economic and human loss. Therefore, it is fundamentally important to design the SHM system, which is proactive in terms of private, public or national interest. In this context, continuous monitoring seems to be an effective solution. *In situ* monitoring, which is a continuous monitoring system, is capable of identifying major differences between vibration-based measurements and environment-based changes. This is one of the important sources of complexities, which actually confuse the data obtained from the sensors, to really work into the application of the measured data towards assessment or control design from the SHM scheme. But, continuous monitoring is expensive and it handles a big volume of data. So, the data communication, data analysis and retrieval can be a sort of challenge in terms of its volume. Certain researchers have also suggested the other alternative for this problem. One of the important alternatives for the above problem is numerical simulation. Numerical structural analysis is also used to predict the health of the structure. It can also avoid complexities that arise from continuous monitoring. For example, continuous monitoring of a bridge is considered. It may involve a lot of complexities including blocking of traffic, conducting expensive static and dynamic load tests, which are essentially cumbersome procedures. Alternatively, the damage status of the deck slab of the bridge can also be detected by analysing eigenfrequency or stiffness degradation. One of the important demerits of this alternate method is that the effects caused by local damage cannot be predicted or detected by this method. There are other specific issues with respect to capturing the time-dependent change in material properties. It is also difficult to capture the time-dependent change in the structural form and the loading pattern. Interestingly, these are the actual sources of uncertainties as well.

1.12.1. *Sources of uncertainties*

(a) *General uncertainties*:

(1) Exact modelling of external load events including its time dependency and space dependency is generally approximated by a set of independent events.
(2) Strength and stiffness degradation with space and time dependence are disregarded.
(3) Measurements of geometric data such as maximum deflection of the deck slab in the case of bridge displacement under dynamic load test are prone to a lot of human errors and inaccuracies.

(b) *Modelling uncertainties*: The structural modelling which indicates the modifications such as construction errors, changes in structural geometry (marine growth, crack propagation, etc.), change in material characteristics due to ageing, physical, chemical and mechanical degradations cannot be captured completely.
(c) *Uncertainties from load variation*: Load can vary with respect to time and space, and it cannot be captured completely.

1.12.2. *Solutions to uncertainties*

The above uncertainties can be handled in three ways:

(1) using random variables;
(2) using fuzziness;
(3) using fuzzy randomness.

1.12.2.1. *Randomness*

The data can be plotted as a typical power spectral density function by considering a probability distribution function and the randomness can be expressed as a PDF function as shown in Figure 1.3.

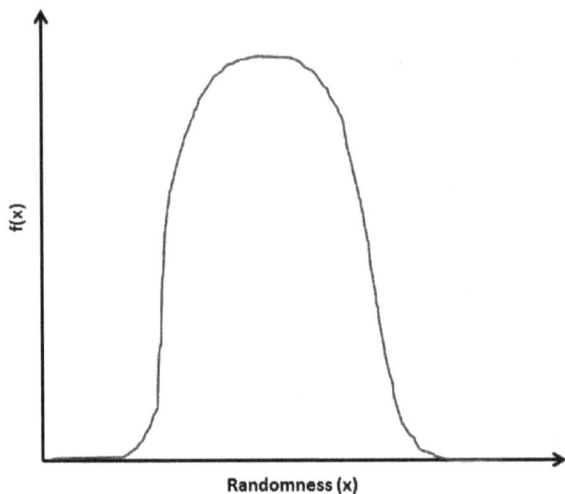

Figure 1.3. Randomness.

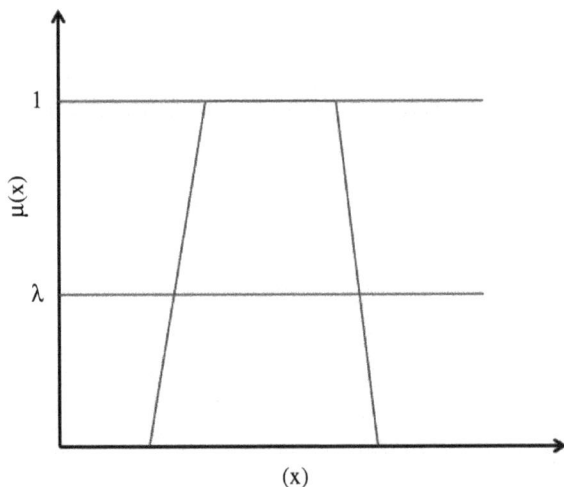

Figure 1.4. Fuzziness.

1.12.2.2. *Fuzziness*

This can be done by reporting data by using a fuzzy logic algorithm which can have a variation modelled typically, as shown in Figure 1.4.

1.12.2.3. *Fuzzy randomness*

This is a combination of fuzziness and random variables which can handle fuzzy randomness, as shown in Figure 1.5, where fuzziness is operated for a particular band and randomness is chosen within the band.

The selection of the model among the above three depends on the availability of the data as these three models are very strongly data-dependent. The quantum of data and quality of data available to represent uncertainty will decide the type of model. For example, if the data are statistically sound, then the parameter can be described once stochastically. But even in this case, appropriate choice of probability distribution will actually affect the results of simulation significantly. On the other hand, if the data of parameters are frequently fragmented and they are not continuously distributed without precision, then fuzzy randomness model is more effective to model this uncertainty.

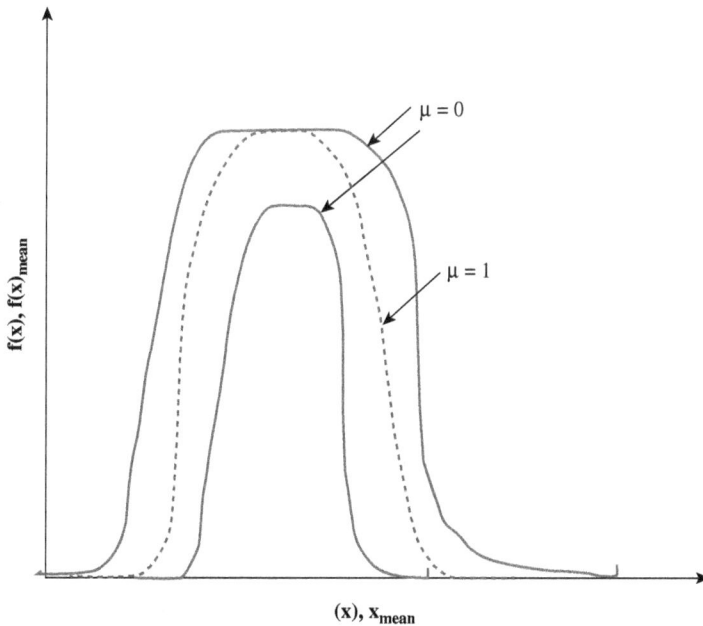

Figure 1.5. Fuzzy randomness.

1.12.3.   *Other classes of uncertainties*

(1) Uncertainties present in the observational data may arise from experimental or numerical analysis. This can be handled by statistical sampling, hypothesis testing and input−output effect analysis methods. These will be useful to characterise the effects caused by the uncertainties on the output of analysis.

(2) There are a few issues related to the advancements in SHM, which can also cause uncertainties. The SHM process, in the present context, is highly advanced. It has got wireless decentralised sensors, which are recent advancements in semiconductor devices and MEMS technology. While they are useful to collect *in situ* data efficiently, they are also capable of establishing communication requirements between the sensors; this makes the decision-making process much faster. However, investigation of the collected data is a critical issue as this needs a faster investigation process. Even though the data are huge in the case of continuous monitoring, modern statistical tools are capable of handling this volume without leaving any residual error. For example, statistical pattern recognition (SPR) can be one of the effective tools. While doing statistical analysis for a set of observations, it is interesting to note that they follow a pattern. Instead of analysing the new set of data every time, this statistical pattern is identified, which makes the process easier. Further, it also simplifies data condensation and feature extraction from the data. SPR plays an important role by offering a major solution to address the uncertainties related to volume of data; this problem is precarious in the case of continuous monitoring of structures.

(3) The most critical issue is data normalisation, which is a process of qualitative separation of data of vibration-based results from that of the data based on environmental conditions; this is done through a statistical prognosis. Even in the stage of statistical prognosis, uncertainty influences service life prediction; time and space dependency characteristics of material resulting from this is very complex. Further, load variations which are also time- and space-dependent add to the complexity.

1.13. Health Monitoring of Ocean Structures

Offshore structures are huge and massive structures, which runs over a large span with heavy mass concentration spread over the entire area. In addition, deep-water platforms need a huge capital investment for their installation itself before it starts its operation to earn revenue. They have many people working onboard during their operation time. Therefore, offshore platforms are too expensive as their downtime could cause a huge financial loss. Most of the offshore platforms in the recent past are planned to operate under unmanned conditions; they are designed to be self-operating and self-producing. In such cases, continuous monitoring is very vital to ensure proper functioning of the platforms. Marine structures like coastal jetties should not be frequently intervened for repair because this could affect the functional value of the system. These structures demand preventive maintenance while the system remains functional. So, structural repairs should be carried out without the shutdown of the system. Most importantly, these structures need to be repaired when they are loaded. In order to understand the response behaviour under such loading conditions, it is important to have a continuous monitoring.

Ocean structures are assigned with special operations like coastal protection, dry-docking, berthing of vessels, coastal defence systems and marine police stations, etc. Safe upkeep of these structures ensures serviceability throughout their life. Four activities that are vital for their upkeep are as follows: (i) condition assessment; (ii) periodic maintenance; (iii) controlled inspection; and (iv) safe operations. Critical factors that influence repair of such structures include the following:

- The structure shall remain in service during repair as a total shutdown is not permitted due to emergency demands that may arise any time.
- The repair methods proposed should be cost-effective and impart long-term solutions as frequent interventions into such structures for repair is not permitted.

- The repair procedures are invoked only under emergency situations and hence continuous monitoring is required to assess the current state of health of the structure to recommend appropriate repair solution.

Having known the fact that visual inspections that are used for routine condition assessment of ocean structures pose serious limitations, recent studies showed their slackness in the desired levels of reliability. It is a common practice to carry out repair of cosmetic type whenever visible damages like cracks, spalling of cover of concrete, chemical deterioration and corrosion are noticed. Inspection helps to identify signs of damage, but correlating them to condition assessment is highly problem-specific and is quite tedious. Unless continuous monitoring is carried out, one will not be able to justify the necessity for any kind of repair in ocean structures. This is due to the fact that these structures cannot be intervened for repair procedures until a thorough reasoning is established.

Some of the major factors influencing accidents in offshore structures are due to poor maintenance, lack of communication between maintenance and operation staff, delay in scheduled maintenance, inadequate maintenance and safety procedures (Chandrasekaran, 2016a,b; Okoh and Haugen, 2013). Damage scenarios are due to various reasons, which include ship impact on the structure, fatigue and damage due to corrosion. Researchers have simulated the damage either by inducing a crack or by removing the member. Experiments are performed on the scaled platform tested in lab to detect the self-induced damages (Begg *et al.*, 1976). In real-time implementation of health monitoring, an attempt to carry out measurements below water line will be expensive and could be highly rational as divers are employed for measurements. Alternatively, measurements below water line can be carried out using underwater sensors with wiring, for which the wiring has to be done at the time of construction of the platform. In such cases, corrosive saltwater environment is an important factor to be considered for the long-term monitoring.

Offshore structures operate under high-risk factor due to the kind of process involved in exploration and production. Apart from being novel and expensive of its kind, their failure may lead to serious environmental pollution and economic loss as well. To avoid such serious consequences, it is better to follow the preventive maintenance approach, which shall depend on continuous monitoring of the structure under time-varying loads. In the case of offshore structures, it is difficult to carry out traditional inspection through NDT or visual inspection as the structure is huge and partially submerged in the water while manual inspection is not possible in all locations. An automated monitoring like SHM using sensors will be an effective tool to assess the status of the platform based on the damage analysis to ensure safe operability.

In the case of offshore structure, as most of the areas are not accessible for measurement, damage assessment scenarios are generally examined through simulated numerical models while a few on the scaled models are analysed through experimental investigations. It is seen as a good practice to analyse changes in response of scaled platforms under different loads during experimental investigations and extrapolate the data for failure analyses or correlate these data with those measured on the real platform. While deploying SHM in offshore structures, certain assumptions and approximations are important for a convenient procedure. They are as follows: (i) varying mass is not linked with the marine growth, equipment and fluid storage; and (ii) variable submergence, leading to change in buoyancy and mass of structural members, is not included and it will alter energy dissipation of the system (Brincker *et al.*, 1995). Based on the studies reported in the recent past, factors that govern the design of monitoring system for offshore platforms are as follows (Loland and Dodds, 1976):

- Sensors should withstand environmental uncertainties.
- Proposed SHM scheme should have financial advantages over the traditional (manual) inspection method.
- Vibration spectrum should remain stable over a period of time.
- Normal sea state and wind excitation shall be used to extract the resonance frequency.
- Above water measurements should be used to identify mode shapes.

1.13.1. *Damages*

Damage is defined as the change in material properties, or the change in geometric characteristics of the system, which adversely affects its current or future performance. This also includes change in boundary conditions and system connections, which can lead to adverse effects due to their degradation. Damage is not meaningful without comparison between two system states: initial and damaged ones. Damage assessment can be simplified by answering the following questions: (i) Is there a damage in the system? (i.e., identification of its existence); (ii) Where is the damage in the system? (i.e., identification of its location); (iii) What kind of damage is present? (i.e., identification of the type of damage); (iv) How severe is the damage? (i.e., evaluation of the extent of damage); and (v) How much of useful life remains? (i.e., prognosis). Answers to the above questions can be readily obtained by using non-destructive tools and by employing non-destructive evaluation methods (NDT/NDE). These tools are very helpful in identifying the damages at the global level. For example, damages on the structure as a whole can be identified but cannot be precisely located at the local level on each member. In the case of reinforced concrete structures (RCC), this problem is more serious due to increased complexities arising from embedment of reinforcement.

While SHM is the process of implementing damage detection strategy for engineering infrastructure, usage monitoring measures input to the structure and its response to these inputs before damage, which can assist in identifying the onset of damage and deterioration. Prognosis is the coupling of the above information of SHM and usage monitoring under given environmental and operation conditions, component- and system-level testing and modelling to estimate the remaining condition and useful service life. One of the recent approaches, which can handle this problem is statistical pattern recognition (SPR). Damages generally initiate at the material level. They are called either defects or flaw. Under certain loading conditions, these damages tend to propagate and they can result in system-level damage. The main concern is not the system-level damage; it is the component-level damage. It is very important to note that the damages do not refer to loss of system functionality. If the system functionality is lost, it is called as failure. Damages prevent the system from performing in its optimal manner. Damage degrades the performance of the system. It does not affect system functionality completely whereas is a total loss of functionality. Damages can be corrected whereas failure needs to be mitigated, where the system has to be reconstructed. So, the purpose of SHM is to avoid failure. Damage cannot be avoided because it is an inherent property of the system which leads to loss of functionality owing to material degradation, excessive loading, excessive deformation, etc. Damage cannot be prevented, but failure can be avoided. Thus, health monitoring will address the failure of the system, which is a total loss in functionality.

Damage to a civil and structural system can occur in two timescales: long-term timescale and short-term timescale. Damages in the long-term timescale can be caused by corrosion and fatigue, while those in the short-term timescale can be caused by impact loads, shock loads and aircraft landing in aviation industries. SHM can be redefined as a process of implementing damage identification strategy. This process involves the following:

• Observation or monitoring on a continuous scale;
• *Assessment based on the extracted data of damage scenarios*: It depends upon the sensitive features identified to quantify damage and the statistical analysis tools which are used to quantify damage. This will help in determining the current state of the structural system. In this process, non-destructive evaluation plays a very important role. It is primarily used to characterise the damage and check for severity when there is prior knowledge of the damage.

1.14. SHM Challenges

Vibration-based damage detection is a very useful and successful tool which is used in civil structures, especially bridges. The outcome of the study is generally modal parameters which are useful as primary features to identify the local damages caused on the deck slab of bridges. They are capable of locating the damage to a larger extent. One of the major concerns in applying SHM to civil infrastructure is the physical size of the structures. In the case of bridges with very long-span deck slabs spread for kilometres, it is very difficult to have sensor network in terms of its reliability, data transmission and its functional robustness. It can be one of the difficult areas to diagnose in terms of the application of SHM on civil infrastructures.

Under the cloud of vibration-based monitoring techniques, one of the basic challenges of SHM is feature selection as this significantly affects stiffness, mass and energy dissipation properties of a system being monitored. Structural damage is a local phenomenon, which may not influence the global response of the structure in lower frequencies, when measured under normal operating conditions; but to ascertain the extent of damage, responses in lower frequencies are equally important. Another important challenge is the training of the SHM system for feature selection and damage identification under unsupervised learning mode. Actually, as data from damaged systems are not available instantaneously, accumulation of damage over wide-varying timescales pose significant challenges. A more intrinsic challenge is the choice of the appropriate sensor system and an accommodative network, which can remain functional even when the structure is expected to get damaged. To be very precise, even upon failure of a sensor, the damage identification algorithms should be capable of adapting to a new (available) alternate network. Researchers recommend the use of self-validating sensors or the use of sensors that report on each other's working conditions. The most important non-technical challenge is to convince the structural system owner that the SHM technology will provide an economic benefit over their current maintenance approaches; convincing regulatory agencies that SHM will provide life-safety benefits is an add-on to this challenge.

1.14.1.   *Challenges in the oil and gas industry*

Oil platforms are generally inaccessible for damage inspection. So, vibration-based techniques have been tried for damage identification in the early 1980s in oil industry. There are some specific issues which make this application a highly challenging technique. The major challenge is that the damage location is not known because majority of the area of the platform is inaccessible for measurement. The most common solution is to simulate the damage scenario using numerical model in a software and examine the severity to interpret the damage. There are major concerns in using vibration-based damage detection in oil platforms. They are as follows:

(1) The machine noise created by the platform interferes with the measured vibration.
(2) Instrument deployment in hostile environment is also a challenge.
(3) A faulty mass representation arises due to marine growth. In vibration-based damage detection, natural frequency of the system is the major parameter, based on which damage is characterised. Natural frequency is a function of stiffness and mass of the system. When there is a faulty mass which arises on the platform due to marine growth, it does not give the damage characterisation exactly as simple as it is applied to other structures.
(4) The varying hydrodynamic mass arises from the fluid storage variation.
(5) Variation arises in the foundation condition in due course of time.
(6) Absence of wave force as exciting force in higher modes also poses challenges.

The above factors or concerns have limited or restrained the use of SHM in the oil industry, especially on oil production platforms. But they are very well and continuously used in ships.

1.14.2. *Challenges in the aviation industry*

In the aviation industry, the vital component for measuring the responses will be on the aircraft. Aircraft are essentially metallic structures, which are designed for specific flight hours. Generally, aircraft are retired from flying once they reach predecided flight hours. Alternatively, if they complete predecided landing cycles, then too, they can be retired. However, if you really do some real-time fatigue analysis and damage assessment of aircraft, especially during landing cycles, then it is possible to extend the flight hours or to preretire them, which improves public safety. One of the methods by which this can be achieved is the use of strain gauges and mechanical strain recorders to measure the stress deviations, particularly during the landing cycles. The second application could be the use of flight data recorder (FDL) using electromechanical mission computer (EMMC). Both methods are very useful in estimating the life of an aircraft and its fitness for 'n' number of flight hours in the future. In both cases, the use of SHM is clearly seen as a major advantage. This can be helpful to decide the suitability or fitness of the aircraft. It is also helpful to modify the design philosophies which are essentially arrived at based on continuous monitoring. There are some anomalies which can be very well explained through the SHM applications if they are in practice. They are as follows:

- The first anomaly is with respect to the statement in aircraft design: "Aircraft geometric configuration is not related to the structural load disbursement." By continuously monitoring the stress values, this assumption can be proved wrong. It is found that the aircraft geometry or configuration makes a significant difference on structural loads.
- The second anomaly states that the "usage of all aircraft in a large fleet averages out with time". This can also be proved to be wrong through continuous monitoring because this is not true based on fatigue assessment. The fatigue damage depends on the actual usage and hence cannot be averaged for a large fleet.
- The third anomaly is based on the maintenance management, which can be planned on the basis of design load spectrum. SHM will definitely help in following the actual measurements of the stress variations. Based on this fact, it was found that average user spectrum is more severe than that of the design spectrum.

Thus, the major anomalies in the case of geometric configuration, average fleet time and in the usage of design load spectrum can be easily understood in a better format when the aircraft is continuously monitored for its performance during landing and take-off processes (Khan *et al.*, 2014).

1.14.3. *Tools of SHM in the aviation industry*

The common tools used for removing the anomalies in the aircraft or aviation industry based upon health monitoring are as follows:

(1) fuzzy pattern recognition;
(2) neural networks;
(3) diffused ultrasonic-waves technique to detect the structural damage present in the unmeasured temporary members;
(4) vibration-based technique;
(5) intelligent parameter varying technique for the location of damage;
(6) novel sensor layout in SHM.

If you look at the maintenance of aircraft structures, one can see the use of health monitoring playing a major role in their successful maintenance. Figure 1.6 shows the parameter contributing to the maintenance of aircraft, where a severe usage pattern is compared with design usage pattern and mild usage pattern. The hatched portion above design usage line clearly shows the safety risk region. The shaded portion below design usage line shows the potential life enhancement region. This figure shows that the severe usage represents the loss of life due to non-SHM. The safety risk region clearly shows the advantage of enhancement of life due to the usage of SHM. Thus, the application of SHM can help in checking the underuse of service life of an aircraft and it can enhance the service life or usage value of aircraft by continuously monitoring its stress distribution levels.

In this context, researchers also recommend passive SHM and active SIIM. Passive SHM deals with the observation of a structure as it evolves. Basically, a physical parameter and its state of evolving as a result of interacting with the environment is examined; tools used could include, e.g., acoustic emission. Active SHM deals with a system where a structure is equipped with both sensors and actuators. This is highly suitable for structures which are unmanned.

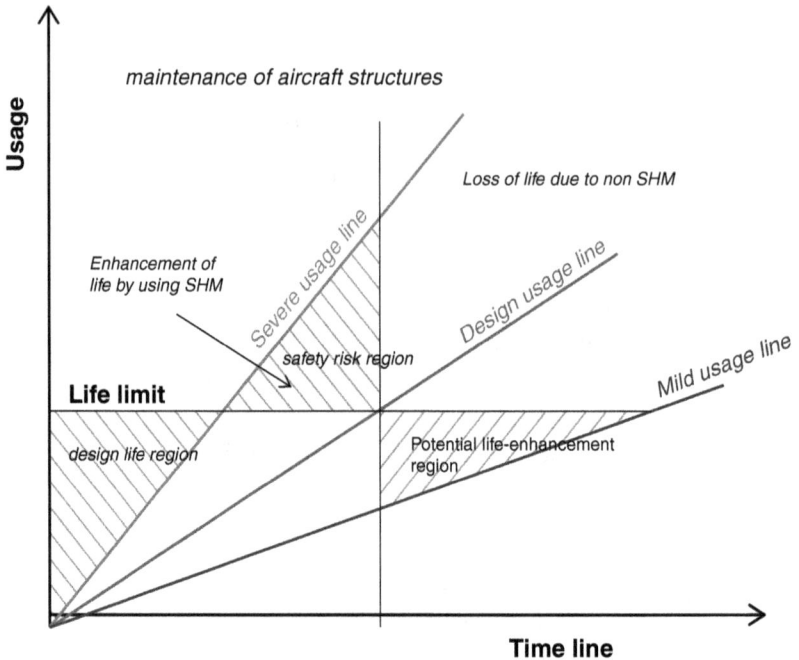

Figure 1.6. Components in the maintenance of aircraft structures.

In such cases, the actuators prompt forces opposite to the structural motion and intuit a recentering capability of the platform or the system under environmental loads. There are many aircraft, which are embedded with the SHM systems. One example is Boeing 787 Dreamliner, which is equipped with embedded sensors for continuous health monitoring. The location of these sensors is very important; generally, they are located in shell fuse legs, lower wing skin and door shutters. These are the places where the probability of damage is relatively high during loading.

1.15.   Successful Deployment of SHM

The basic components of SHM are sensors, actuators, smart structures, smart materials, computational systems, signal processing and statistical models, as shown in Figure 1.7. There is a small overlay of

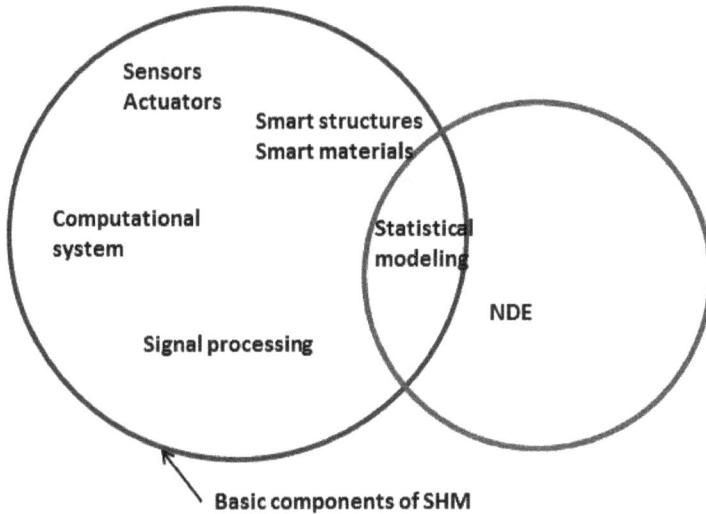

Figure 1.7. Basic components of SHM.

NDE with basic components of the system. NDE is a vital component integrally connected to SHM.

1.15.1. *Recently deployed NDE techniques*

(i) **HELP:** This refers to hybrid electromagnetic performing layer. This method is an alternative to a fully integrated electromagnetic technique which is quite expensive. In this method, by embedding the network of conductors in the material or by bonding the network of conductors on the internal surface of the structure, health monitoring is carried out. In this process, a grid which is a sensitive magnetic field is created. This field is created by an external electromagnetic antenna which crosses the structures. This is made of conductive composites, e.g., epoxy composites, carbon composites, etc.

(ii) **Ultrasonic vibrothermography:** In this case, lamb waves are used to generate ultrasound. The embedded piezoelectric patch monitors the surface thermal field with the help of camera. This field is produced by the interaction of lamb waves with the structure having defects (crack, fissure and delamination). This technique is very helpful to study delamination in composites.

(iii) **Lock-in shearographic imaging of ultrasound:** Shearographic imaging is created or generated by piezoelectric patch, which is generally embedded on the surface of the structure.

The above methods are very advanced and helpful in assessing the actual control and health of the structural system to a larger extent.

1.16.   Axioms Used in the SHM Process

There are some common axioms used in the SHM process which are very useful and interesting.

Axiom I: All materials have a few inherent flaws or defects.

Axiom II: The assessment of damage requires comparison of the two systems. Damage assessment is always relative (with respect to another system). It is always on the comparison mode.

Axiom III: Unsupervised learning mode can be helpful in the identification and location of damage.

Axiom IV: Sensors cannot measure damage. The data collected or acquired are to be processed to extract featured values which can then be used to detect or quantify damage. Thus, the vital part is the signal processing of collected data and the statistical classification of the data to convert the sensor data into damage information.

Axiom V: The more sensitive a measurement is to damage, the more sensitive it is to change in environment and operational conditions. It means that there is a very high possibility of noise mixture with the damage data. Hence, it is important that one should intelligently extract the featured information from the recorded or acquired data.

Axiom VI: It is related to the sensing system. The SHM sensing system strongly depends on length (period) and timescale associated with the damage initiation and evaluation. For example, if a damage is a long-term phenomenon, then the damage propagation in terms of its timescale can be lost if not handled with appropriate sensing system.

Axiom VII: There is a strong correlation between sensitivity and damage. Therefore, the algorithm used to extract featured information based on which damage will be quantified should be carefully chosen and it should be free from noise reflection capability.

Axiom VIII: The size of the damage that can be detected from an SHM system changes with the dynamics of the structural system. It is inversely proportional to the frequency range of the exciting force.

1.17. Real-Time Monitoring of Buildings Under Seismic Excitation

This is a case study which is applied to a set of buildings post earthquake at San Francisco (Çelebi *et al.*, 2004). It is studied after the 6.0-magnitude earthquake at San Andreas, the peak ground acceleration of this earthquake exceeded 0.25 g, which is expected to cause considerable damage to different types of buildings. The objective of this study is that the owner of the buildings wanted to assess the safety of occupancy after the earthquake. A real-time monitoring was deployed to carry out this assessment. The requirements of this scheme are as follows:

(1) The systems must facilitate a rapid assessment of building integrity based on which the occupant safety can be declared following the earthquake.
(2) The SHM system must provide data like drift ratio, which is related to the earthquake damage. They can be used as indices to quantify the occupancy level of the system post earthquake.
(3) SHM should also have the capability to deliver data within few minutes after the occurrence of earthquake.

The SHM system recorded the following:

• The SHM system waits for an event.
• Once the event occurs, it produces a low-amplitude data in real-time analysis and starts making assessment of the data.

- Data provided by the system is very useful for post-seismic assessment.
- Based on the structural type and the damage assessment made, condition of occupancy post earthquake was to be declared based on different guidelines (e.g., FEMA).

1.18.   SHM Issues in Concrete Structures

Concrete is one of the favourable construction materials for offshore structures and civil engineering structures. Nevertheless, concrete has shown greater advantages of structural use because of its strength, availability, application criteria and different structural forms which can be amendable using concrete as a construction material. However, there are some specific issues in health monitoring of concrete structures, which require a fundamental understanding. There are specific kinds of NDE methods which are very useful and particularly powerful when applied to concrete structures. In general, issues arise due to the special kind of problems in concrete structures. They are affected by a variety of issues. Fundamentally being heterogeneous material, issues are further complicated. When we talk about strength degradation, concrete has several ways by which strength is degraded. They are chemical degradation, physical degradation and mechanical degradation.

Examples of problems in concrete are related to the following: (i) chloride penetration; (ii) sulphate attack; (iii) carbonation; (iv) freeze–thaw cycles; (v) shrinkage of concrete; and (vi) issues related to mechanical loading on concrete structures. Chemical degradation includes corrosion of reinforcement, chloride penetration, carbonation, leaching of concrete, concrete under acid attack, sulphate attack and alkali aggregate attack. Physical degradation includes temperature variation and associated thermal expansion or contraction. There can be issues causing physical degradation also arising from the variation of relative humidity which is very important in coastal structures and issues associated with drying shrinkage and wetting expansion. Frost attack, wear and tear and abrasion can also cause physical degradation. Mechanical degradation arises essentially from externally applied load which can sometimes cause overload or impact load. There can be issues due to fatigue loads, differential settlement of foundation of concrete structures and seismic activity.

1.18.1. *Influence of degradation on concrete*

Concrete degradation can alter the properties significantly as follows:

- These degradations can alter porosity and permeability of concrete.
- It can further initiate or aggravate different material flaws, such as scaling, spalling, swelling, debonding, cracking, disintegration of concrete, etc.
- The degradations can also cause impairment in water tightness of concrete members. Especially reservoir structures, dams, overhead water tanks can also be affected very severely by this condition.
- It can ultimately reduce the load-carrying capacity of the member.

1.18.2. *Major challenges and solutions*

A major challenge in monitoring the health of concrete structures is that the damage under different deterioration processes accumulates at different rates. The timescale variation of these degradations is different. They get integrated and mixed to alter the behaviour of concrete. Therefore, there will be a multi-physics degradation process which needs a special type of analysis that can account for different timescales in different processes of degradation. If a numerical analysis is required to be carried out, then the governing differential equation (time-variant and space-variant) should account for the coupled physical and chemical process dependency. It should characterise the following:

(1) mass energy balance;
(2) thermodynamic and chemical equilibrium of the coupled heat conduction, ionic diffusion, moisture transport phenomenon and the corresponding chemical reaction.

Therefore, it is very complex to analyse this numerically. The major factor that contributes to the degradation of concrete should be identified. Interestingly, ordinary concrete possesses a high porosity and low permeability. Now, the interconnected pores or micropores and microcracks in concrete contribute to the permeability. Therefore, this makes concrete more vulnerable for deterioration. It is actually the rheology and crack structure of concrete which makes it complex. So, health monitoring of concrete is not a physical process. It is also not an electronic process where one can simply measure the strain values, displacements and deformations. It is also not purely a chemical process because it also contributes from other sources which are physical and mechanical. So, health monitoring of concrete essentially becomes multi-physics-dimensional problem.

1.19. Non-destructive Evaluation

There are many non-destructive evaluation (NDE) methods which are exclusively available for concrete structures and they can reasonably give good health condition for concrete structures. Some of the NDE methods suggested by Clayton (2014) are as follows:

(1) shear wave ultrasound;
(2) ground penetration radar;
(3) impact echo analysis;
(4) ultrasonic surface wave analysis;
(5) ultrasonic tomography.

In addition, for large volume structures, one can also use full-field imaging techniques, e.g., gravity-based platforms, nuclear reactors, etc.

1.19.1.   *Full-field imaging techniques*

These techniques are useful for concrete structures, which essentially can be applicable to large volume structures.

(1) **Infrared imaging:** It tracks the thermal load path in a material, travelled longitudinally over a period of time. The onset changes in the load path changes the composition of the material which is an indication of the mechanical damage caused to the material. This method can also be combined with acoustic source or stand-off acoustic sound pressure technique to quantify the extension of damage. In this case, material is insonified with acoustic source and the full-field vibrothermograph measurements are recorded to characterise the material.

(2) **Measurement of the thermal response under an applied uniform heat flux:** Thermal gradient in the material is measured and analysed to identify the non-uniform material composition which essentially arises from the material defect that can then be characterised. There is another method which can be useful in large volume concrete structures.

(3) **Digital image correlation (DIC) technique:** This is useful to detect microcracking in the chopped fibre glass compressive moulded parts. DIC image shows principal strains in the damaged regions where cracks are formed. This method is useful to detect localised residual stresses, which are caused in the material upon removal of load. This can also be used to track the strain variations that occur under temperature variations.

(4) **Velocimetry:** This method is useful to detect the sub-surface nonlinearity caused due to material damage. For example, when a composite structure is subjected to ambient vibration, changes in strain variation can be analysed to detect the damage. In such cases, the damage indices quantify the degree of nonlinear stiffness and nonlinear damping, which are generally observed locally at each measured point on a grid of the member.

1.20. Sensor Performance and DAQ

Quality of data in health monitoring system depends on the performance of the sensors. Some common factors that govern are data format, precision and accuracy, linearity of data, dynamic range, cross-talk, durability, maintainability, redundancy, calibration and its cost. Structural health monitoring involves detection and tracking. While the first step is to make the sensor reading correlated with the sensitivity of damage, tracking involves establishing the relationship between the damage features and the damage levels. In general, there are two approaches in which the SHM system is developed. The most common strategy for developing the sensor network for SHM is to deploy an array of sensor network with the commercially available components. The excitation of the structure is limited to the range of frequency of this array of sensors. Physical quantities are measured without any definition of the damage that has to be detected, with an assumption that these measured data will be sensitive to the damage (Farrar *et al.*, 1994). This is based on the assumption that damaged and undamaged structures are subjected to a similar kind of excitation. Such strategy is deployed in real time, which will measure the data and analyze the data for damage-sensitive features. The alternate strategy is to quantify the damage through some process before developing the sensing system. Based on the available numerical simulation results, damage location and type of sensors are chosen. The extraction of damage features and statistical pattern recognition will be a part of the data analysis, which will be vital in the development of the data acquisition system (Flynn, 2010). Additional requirements are updated based on the changing environmental and operational conditions. The latter ones with the initial prediction about the damage by numerical simulations improve the probability of damage detection.

Types of data that need to be acquired should be defined to design the sensor network system. Two major types of data are as follows: (i) kinematic quantities and (ii) environmental quantities. Kinematic quantities include displacement, velocity, acceleration and strain. Traditional types of sensors are used to measure these dynamic responses. Accelerometers, displacement transducers and force transducers like load cells are some of the sensors used in SHM. Environmental quantities include temperature, pressure and moisture content. These parameters not only affect the damage level of the system but also have an impact on the operation of the sensors.

1.21. Need for Wireless Sensor Networking

SHM is a typical field in which the application of wireless sensor network (WSN) is useful both to measure damages and for online monitoring. Due to the reducing price and advancements in recent technologies of sensor networking, it is now easy, simple and affordable to have WSNs in lieu of the traditional wired SHM. Wireless SHM reduces the system cost and the installation time. In addition, it eradicates the installation of lengthy cables and thereby reduced complexities are involved in their laying and in-service maintenance. While the wired system depends on the centralised server, wireless nodes do not rely upon a central server. They convert the measured data into a digitised form and transmit them directly. Wireless SHM makes online monitoring more simple with low-cost computing processor. Recent innovations in the wireless SHM leads to the migration of computational power from the centralised data acquisition system to the sensor nodes.

1.22. Critical Issues in SHM

Though SHM has a wide variety of applications in various fields, there a few critical issues as follows:

- Uncertainties can arise from parametric data owing to the physical experiment and numerical simulation output.
- The imperfect knowledge of the control parameters of the physical experiment and numerical simulation also adds to the imperfect knowledge on the input to numerical model.
- Uncertainties can also arise from stochastic equations of motion, environmental variations, measurement errors which are human-based, discretisation and numerical errors.
- Uncertainties can also arise from the probability density functions of specific probability distributions. Probability density functions can handle problems related to uncertainties using random theory. So, one can choose a specific type of distribution to include all possible values of the variable. Other methods of handling this uncertainty are as follows:

(1) Dempster–Shafer theory of possibility and belief;
(2) theory of fuzzy set;
(3) information gap theory;
(4) convex model of uncertainty.

- The simpler way to address this uncertainty is the Monte Carlo technique. It is an idea towards randomly picking values of a parameter such that the histogram of the chosen values approximates the probability density function. Subsequently, the computational mode is analysed or evaluated at each point sampled in the input parameter space.
- Although many methods exist for the identification and location of damage, no single method solves all problems in all structures.
- The SHM techniques and process have damage-related sensitivities. While a sensitive technique produces false-positive results, less sensitive technique shall produce false-negative results.
- More nonlinear parameters are used in the SHM process, making the process more complex.
- While most of the traditional detection methods are based on appreciable reduction in the rigidity of structural element, quantification of damage and life prediction is more complex. This is due to the fact that correlating reduction in rigidity to decrease in strength is a difficult task.
- Although the SHM process has shown a lot of errors that may arise from experimental investigations, incompleteness and environment-related problems, statistical methods are very effective in handling them.
- Statistical pattern recognition (SPR) enables reduction in the number of sensors to be deployed for SHM but still requires advanced research to ensure correlated results that are authentic and safe.

Chapter 2

Structural Health Monitoring: Detailed Perspective

This chapter deals with details of structural health monitoring methods and techniques. Damage detection methods that are commonly deployed for civil engineering infrastructure are explained in detail. Damage identification using various methods is also discussed, while a comparison is made between these methods to highlight their applicability issues. Non-destruction evaluation methods with their application issues are also discussed briefly.

2.1. Focus of Structural Health Monitoring

Structural health monitoring (SHM) is mandatory to assess old structures in an effective manner. This may be required for various purposes: (i) repair and rehabilitation; and (or) (ii) to assess the cost of rebuilding. SHM needs to focus on the following four stages:

(1) inspection;
(2) investigations including both experimental (*in situ*) or experimental (lab-scale) and (or) analytical (either scaled model or prototype);
(3) monitoring;
(4) evaluation and assessment.

2.2.   Glossary

Glossary of terms that are commonly referred in SHM are discussed:

Ambient vibration test: It is a vibration test, which is carried out for dynamic tests in SHM where the structure is excited by wave, wind, traffic loads or any other human activities under normal conditions.

Assessment: It is defined as a validation of structural conditions.

Continuous monitoring: It is usually carried out on a continuous basis to determine any detrimental changes in the characteristics of the structure.

Damage: This refers to any change in the health of the structure in terms of its conditions, which decreases its performance.

Defect: This refers to a condition-related deficiency, which is user-defined.

Evaluation: It is a process through which the actual load carrying capacity of the structure is determined.

Inspection: It is a non-destructive examination, which is carried out to detect defects in the (old or new) structural system.

Load effect: It refers to the consequence on the structural member caused due to loads and forces. It also refers to the change in the geometric system of the structure caused by the loads and forces.

Long-term monitoring: It is a process of periodic or continuous monitoring, which is carried out over a long period of time (several years).

Periodic monitoring: This refers to non-continuous (intermittent) monitoring, which is carried out to identify (and quantify) any significant change or detrimental damage on the (old or new) structural system.

2.3.   Necessity of Visual Inspection

It is important that infrastructure projects, such as bridges, tunnels, retaining walls, dams and offshore structures, are generally subjected to over-usage in terms of their service life. One of the main reasons of their over-usage beyond their intended service life is that rebuilding these structures is almost impossible. In addition, the cost of interrupting public life and huge investment of public fund will not essentially recommend their reconstruction. Specific problems associated with such structures are as follows: over-ageing; change in pattern of usage over a period of time; non-compliance with the current code provisions; significant reduction in strength and deterioration of geometric form; heritage value associated with the structure; and non-availability of construction material to match with the properties of that of the old ones. It is therefore imperative to know the current state of health of such structures so that their service life can be prolonged in the interest of public safety. SHM is therefore necessary and possibly the only way to assess their present condition. Further, one can also notice that the critical combination of facts that demand health monitoring are, namely, increased loads, poor maintenance and inadequacy of strength in terms of current code compliance.

Therefore, all public structures demand SHM as an inadvertent process. Most commonly used method to monitor public buildings is through visual inspection in periodic intervals carried out by the maintenance personnel. This is a common practice in public buildings in most of the developed countries such that the period of interval between successive visual inspections can be based on several factors: (i) importance of such buildings; (ii) rate of deterioration; (iii) challenge of public safety; and (iv) budgetary requirements. Usually, the period of interval is about 3–5 years. Visual inspection has certain drawbacks as follows:

- Structural deficiencies during visual inspection can be detected only if those surfaces are accessible. For example, in case of offshore structures, marine growth is a very important natural barrier obstructing the visual inspection significantly. The legs of the platform will be completely covered by marine growth and it will be difficult to know the exact condition of the members; information related to the formation of the marine deposits is also scarce.

- A long gap between the periodic inspections can reduce safety because structural degradation process, if it occurs faster than that of the periodic inspection, will become unnoticeable.

2.4. Objectives of SHM Methods

SHM is a scheme that provides information on demand about any significant change or damage or defect that occurs in the structure. So, essentially, SHM is assessment or at least detection of defect or damage, which has the following objectives:

- Structural phenomena, such as corrosion, cracking, delamination, settlement effects, should be inspected and investigated.
- Time strategy, such as continuous monitoring, periodic or triggered monitoring, should be advised based on the nature of the defect and the type of the structure.
- Condition of the phenomenon, whether it is local or global, should be observed.
- Load effects caused on the structures should be also repeated.
- Evaluation methods should include cause of failure and consequences of failure with respect to degree of severity should be reported. This should cover problems related to structural geometry, material degradation, load data, etc.

For example, let us consider the deck of a bridge as shown in Figure 2.1. It is a two-lane traffic bridge supported by columns and piers.

The parameters to be observed in the bridge in terms of health monitoring are as follows: (i) vibration of the deck slab; (ii) excessive deformation of the deck; (iii) formation of cracks or measurement of crack widths; (iv) corrosion of reinforcement in the deck slab; and (v) settlement of the foundation.

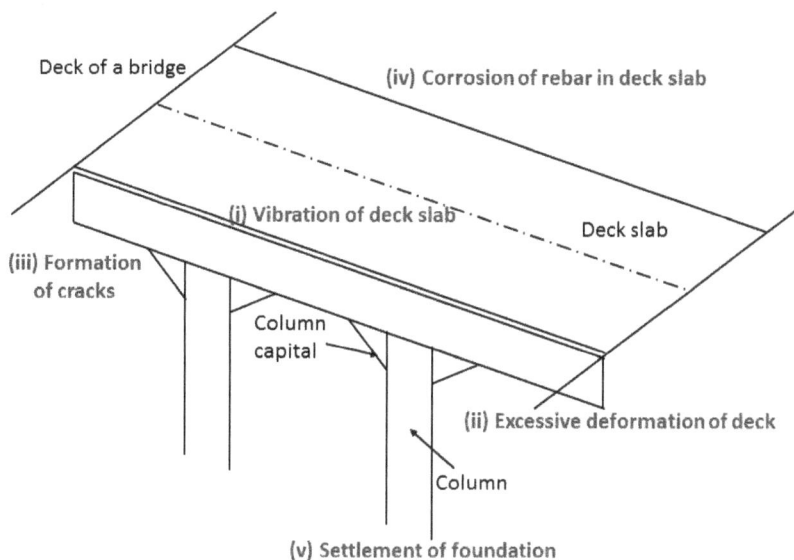

Figure 2.1. SHM parameters for a bridge deck.

2.5. Monitoring

Monitoring can be classified as short-term monitoring (STM), long-term monitoring (LTM) and triggered monitoring based on the period of monitoring.

2.5.1. *Short-term monitoring*

This can be useful for structures that are to be examined only at specific points of time. For example, in a road or a railway bridge, monitoring is required only when there is a heavy traffic. STM can also be done if visual inspection shows a definite damage. In such cases, STM is carried out to validate and collect more details about the damage. Most of the sensors used in STM are not robust and unable to sustain long, periodic observations. Due to this reason, sensors are used in STM only for a specific period of time. They are generally used in "on–off" mode. It is interesting to know that STM, if repeated at periodic intervals, can be a substitute for periodic or LTM.

2.5.2. Long-term monitoring

This can be useful when the period of monitoring is very large (Mufti *et al.*, 2016). LTM is carried out over the entire life of the structure. Specific conditions under which LTM is carried out are as follows: (i) structures that show slow change in loading, such as gradual change in temperature; and (ii) prediction of effect of natural hazards, such as the effects of earthquake, flood, hurricane, etc.

2.5.3. Triggered monitoring

This is carried out when data collection is initiated by a specific event or when a parameter exceeds a pre-set threshold value. In triggered monitoring, sampling interval depends on the dynamic nature of the studied phenomenon, for example, monitoring the vibration when a train pass a railway bridge. Graphically, this can be expressed as shown in Figure 2.2.

2.6. Local and Global Monitoring

Local monitoring refers to the observation of load phenomenon and other local effects, such as crack initiation, crack propagation, extraneous strain, etc. Generally, local monitoring is carried out using

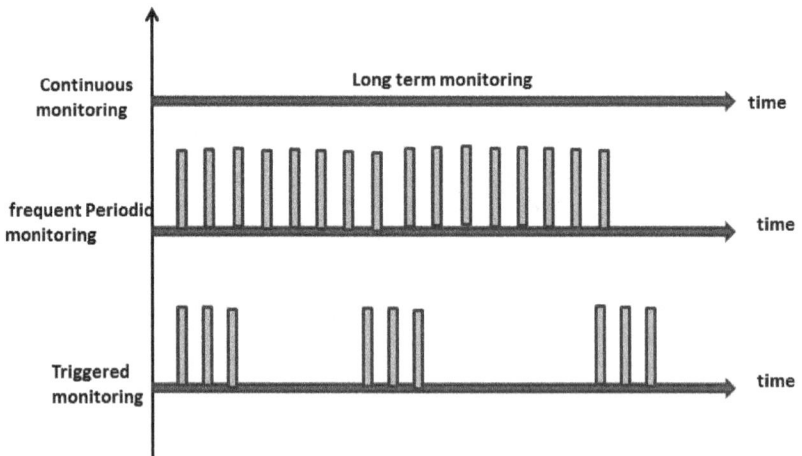

Figure 2.2. Types of monitoring.

non-destructive tests. Local monitoring is useful in identifying (and quantifying) the severity of damage. Mostly, it is very effective if carried out in lab scale as the parameters can be controlled effectively. For example, while modelling the girder of a deck of the bridge under traffic load (rolling loads), speed of the vehicle and the load movement need to be controlled in the lab scale.

Global monitoring will focus to determine deformations (or large displacements) of the structure under excessive loads when it is vibrating. Usually, modal parameters, such are frequency, mode shape and damping estimates, are measured. Subsequently, they are correlated with the analytical studies to identify (or quantify) the damage of the structural system.

2.7. Static and Dynamic Monitoring

Variations of parameters, such as deflection, inclination (in case of bridge piers), settlement of foundation, crack width, corrosion of reinforcement, which are essentially slow-varying process, need to be carefully monitored. Sometimes, their effects cannot be monitored under static loads. Simultaneously, changes can also be caused due to environmental conditions such as temperature variations, humidity, wind force, wind direction, presence of current, wave force, wave height, wave direction, etc. Both these sets of effects put together can sometimes exhibit a quasi-static behaviour, which is neither purely dynamic nor completely static. Therefore, while monitoring these parameters, one should measure only the peak value, which is observed over a long period of time. This is referred as static monitoring.

Dynamic monitoring is carried out with a higher sampling rate. Dynamic monitoring is used to obtain variations in structural characterisations under dynamic load effects.

2.8. Types of Monitoring

There are varieties of monitoring systems and methods, which can be deployed for various purposes. This includes a variety of sensors that

Table 2.1. Monitoring and sensor applications.

Phenomenon	Monitoring strategies	Sensor types
Foundation settlement	• Local • Continuous • Static • Long term	• Linear voltage differential transformer (LVDT) • Lasers • Hydrostatic liquid system • All combinations
Displacement	• Global • Short term • Long term • Periodic or triggered	• LVDT • Non-contact type lasers • GPS
Inclination and rotation	• Local • Short term • Long term • Continuous	• Inclinometer (uniaxial or biaxial)
Crack detection	• Global • Dynamic	• Fibre optic sensors
Crack width	• Local • Periodic • Static	• LVDT • Crack sensors
Vibration	• Global • Short term • Periodic • Dynamic	• Accelerometers (uniaxial/biaxial/triaxial)
Corrosion	• Local • Continuous • Static	• Scanning sensors • Embedded sensors

are used for application of monitoring in the infrastructure industry. They are listed in Table 2.1.

2.9.   Data Evaluation and Assessment

Once data are collected from the sensors, they need to be processed to evaluate the condition of the structure. One of the most common methods of evaluation is using probabilistic tools. The performance of the structure would need to be upgraded, which is one of the important outcomes of assessment of the structure under health monitoring. Under economic constraints, if the revised design of the structure shows higher safety, then one should check this assessment using reliability tools; reliability-based calculation and classification are also important in such cases.

2.9.1. *Reliability*

For a satisfactory performance of the structure, the following condition must be satisfied:

$$R \geq S,$$

where R is the resistance of the structure and S is the load effect on the structure. To determine the load effect on the structure, distribution of loads, in terms of its location, intensity, time and space dependence, direction, etc. are required. These are time-dependant variations and their relationship should be known beforehand to estimate their values. This can be readily obtained from a continuous monitoring data. Subsequently, target reliability index (β), which is useful in assessing the condition of the structure, is estimated. Table 2.2 as recommended by the code (ISO:13822) is the basis for design and structural assessment of existing structures.

Table 2.2. Reliability index (ISO:13822).

Limit state	Reliability index	Reference period
Serviceability		To calculate the
(i) Reversible	0	remaining service life of
(ii) irreversible	1.5	the structure
Fatigue		To calculate the
(i) **can be inspected**	2.3	remaining service life of
(ii) cannot be inspected	3.1	the structure
Ultimate		To design for service
(i) **very low**		life of the structure
consequence of		(about 50 years)
failure	2.3	
(ii) low	3.2	
(iii) medium	3.8	
(iv) high	4.3	

2.9.2.   *Target reliability index (β) and probability of failure (P_f)*

The relationship between the target reliability index and probability of failure is given by

$$\beta = \phi^{-1}(P_f).$$

The relationship is also given in Table 2.3. The table also indicates a safety class, which is introduced depending upon the reliability index.

In general, if the probability of failure or reliability index can be computed, then this can be also verified from the assessment of the structure, which is actually collected from the continuous monitoring data. The probability of failure is given by the following relationship:

$$P_f = p(R - S \geq 0),$$

where (R, S) are stochastic variables. But, as the structures within same classifications are designed for equal loads with different materials, the estimate of probability of failure needs to be modified. Therefore, design codes recommend partial coefficients, both for the load effects and the material. The design value (f_d) will be based on both the partial factors for the material and load effects to cover the modelling uncertainties and error in load estimates. This is given by the following relationship:

$$f_d = \frac{f_k}{\gamma_m \gamma_n} \frac{k}{\eta},$$

where f_k is the stress capacity, which is reduced by the factor γ_m and γ_n, k is the load capacity factor, η refers to the value accounting for model uncertainties and effects accounting for scaling up the lab scale results to a full-scale structure.

Table 2.3. Probability of failure and reliability index.

P_f	10^{-1}	10^{-2}	10^{-3}	10^{-4}	10^{-5}	10^{-6}	10^{-7}
β	1.3	2.3	3.1	3.7	4.2	4.7	5.2
Safety clause				1	2	3	—

2.10.   Damage Detection

When a structure vibrates under the influence of any load, either ambient or service load for which the structure is designed, then the dynamic measurements can be used to characterise the structural condition at any instant of time. It is necessary to locate the damage if it is present within the system, which can be assessed by comparing the structural characteristics at pre- and post-damaged states. It is not necessary that SHM should be carried out only when damage is perceived. On the other hand, structural health can be assessed by comparing the vibration characteristics before and after the damage, even when the structure is in its healthy state. However, one should impose SHM to obtain vibration characteristics before any perceived damage. Even if all structures are designed to cater to the varying dynamic characteristics to avoid damage, SHM is still necessary. This is due to the fact that dynamic characteristics vary significantly with the following:

- changes in loading pattern, invoking redistribution of load;
- ageing of the material or material degradation, causing significant changes in in mass and stiffness of the structure;
- changes in support conditions with period of time.

The above parameters alter the dynamic characteristics of the structure to be different from that of their design state. It is therefore necessary to periodically update vibration characteristics of all the structures. This is vital for at least structures of strategic importance as their health can be assessed readily based on the comparison between their pre- and post-damaged states.

2.10.1.   *Static-based detection*

In SHM, one of the foremost steps is damage identification, which is commonly assessed through dead load distribution. The basic hypothesis is that the dead load of the structural system will get redistributed automatically when a damage occurs in the system. In such cases, stress and strain variations occurred due to dead load redistribution are measured and used as input to identify the damage.

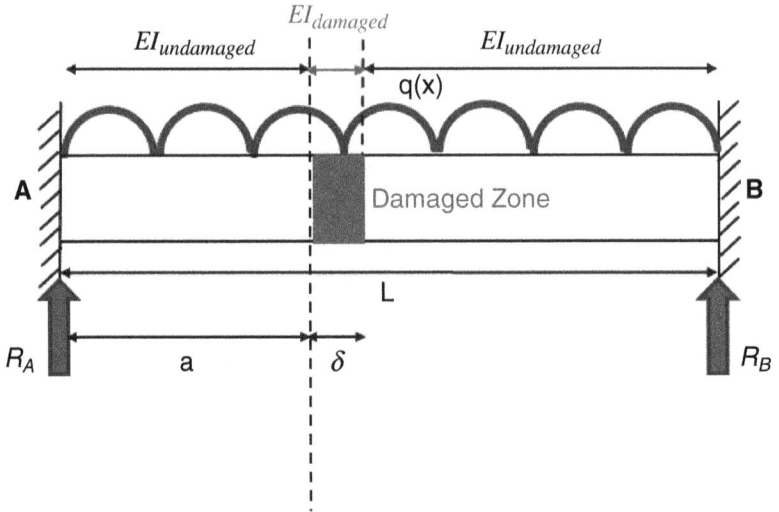

Figure 2.3. Fixed beam with postulated damage.

This is commonly referred to as the static load test. For example, let us consider a fixed beam shown in Figure 2.3. The beam is subjected to some loading pattern for its entire length, which is a function of x, where x is measured positive towards the length of the member while y is measured positive upwards. A moment is applied at the left end 'A' of the fixed beam. Let us perceive that there is a damage at a distance 'a', measured from the left end support 'A'. This will cause variation in the flexural rigidity (EI) of the member in the damaged zone. Let the damage region be of span δ, as shown in Figure 2.3.

Moment at any section $x-x$ is given by the following relationship:

$$M_x = +R_A(x) - \frac{qx^2}{2} + M_A.$$

Let EI_{und} and EI_{dam} be modulus of rigidity of the undamaged and damaged part of the beam, respectively. From the beam theory, the following relationship is well-established:

$$EI \frac{d^2y}{dx^2} = M_x.$$

Integrating the above equation once, we get

$$EI\frac{dy}{dx} = \frac{d\,(M_x)}{dx} + C_1.$$

Further integrating, we get

$$EIy = \frac{d^2(M_x)}{dx^2} + C_1 x + C_2.$$

Substituting for M_x, we get the following relationship:

$$EI\frac{dy}{dx} = R_A\left(\frac{x^2}{2}\right) - \frac{qx^3}{6} + M_A x + C_1$$

$$\Leftrightarrow EIy = R_A\left(\frac{x^3}{6}\right) - \frac{qx^4}{24} + \frac{M_A x^2}{2} + C_1 x + C_2.$$

There are different regions in the beam for which the above equations should be separately written.

(a) *Region $0 \le x \le a$*

$$EI_{\text{und}}y_1 = R_A\left(\frac{x^3}{6}\right) - \frac{qx^4}{24} + \frac{M_A x^2}{2} + C_1 x + C_2.$$

(b) *Region $a \le x \le a + \delta$*

$$EI_{\text{dam}}y_2 = R_A\left(\frac{x^3}{6}\right) - \frac{qx^4}{24} + \frac{M_A x^2}{2} + C_3 x + C_4.$$

(c) *Region $a + \delta \le x \le l$*

$$EI_{\text{und}}y_3 = R_A\left(\frac{x^3}{6}\right) - \frac{qx^4}{24} + \frac{M_A x^2}{2} + C_5 x + C_6.$$

The boundary conditions to solve the above equations are as follows:

(i) $y_1(0) = 0, y_2(0) = 0, y_3(0) = 0$;
(ii) $y_1(a) = y_2(a)$;
(iii) $y_2(a + \delta) = y_3(a + \delta)$;
(iv) $E(x) = \frac{M_x y}{EI_{\text{und}}}$;
(v) $E(x) = \frac{M_x y}{EI_{\text{dam}}}$.

Subsequently, one can establish the damage severity index as follows:

$$\alpha = \frac{EI_{und} - EI_{dam}}{EI_{und}}, \quad 0 < \alpha < 1.$$

Now, the problem is reduced to minimise the function and is given by

$$f(\alpha, \delta, a) = \sum_{j=1}^{k} \left| \frac{\Delta \varepsilon_j^t - \Delta \varepsilon_j^m}{\Delta \varepsilon_j^m} \right|, \quad 0 \le a + \delta \le L, 0 \le \alpha \le 1, \delta \le L.$$

Beam is divided into small length such that following relationship holds good:

$$l = \left(\frac{L}{n} \right),$$

$$\delta = l = \left(\frac{L}{n} \right),$$

$$a = a_i = \delta_{(i-1)} \quad \text{for } i = 1, 2, \ldots, n.$$

Hence, the problem is reduced to minimising the following:

$$f(\alpha, a_i) = \sum_{j=1}^{k} \left| \frac{\Delta \varepsilon_j^t - \Delta \varepsilon_j^n}{\Delta \varepsilon_j^n} \right|,$$

subjected to $a_i = 1, 2, \ldots, n$ for $0 \le \alpha \le 1$.

2.10.2. *Vibration-based detection*

The hypothesis behind this method is that the structural damage can be characterised by local modification of stiffness; change in stiffness in turn affects the modal parameter. In what follows, the procedure is discussed:

- Member will be subject to an external excitation. This excitation can be a forced vibration, which is possible in a lab scale. It is not possible to create a forced vibration system in a real-time scale unless it is automatically created by external loads. Alternatively, it can be an ambient vibration under natural loading cases.

- Under this condition, modal parameters are estimated from the vibration data.
- These parameters are used as input for damage detection.

Let us consider changes in modal parameter as Δ_v, stiffness reduction factor (SRF) as α, weightage of each term in the stiffness matrix of the member as $\overline{w}$, analytical data be represented by A and experimental data represented by B.

Damage identification can be done as follows:

$$J = \|\overline{w}\{\Delta v^{\text{analy}}(\alpha) - \Delta v^{\text{exp}}\}\|^2.$$

The problem is to minimise the above function J subject to the condition that $0 \le \alpha \le 1$ is valid, which is the stiffness reduction factor:

$$J = \{\Delta v^{\text{analy}}(\alpha) - \Delta v^{\text{exp}}\}^T (\overline{w})^2 \{\Delta v^{\text{analy}}(\alpha) - \Delta v^{\text{exp}}\}.$$

Damage detection and quantification can be assessed from the objective function as follows:

$$J = \sum_{i=1}^{nm} \overline{w}_\lambda^2 \left[\left(\frac{\lambda_i(\alpha) - \lambda_i}{\lambda_i} \right)_{\text{analy}} - \left(\frac{\lambda_i^{\text{damaged}} - \lambda_i^{\text{undamaged}}}{\lambda_i^{\text{undamaged}}} \right)^{\text{exp}} \right],$$

where the number of measured modes in the analysis is indicated as 'nm' and λ_i is the ith eigenvalue. In order to trace the changes in the mode shape, the following relationship can be used:

$$J = \sum_{i=1}^{nm} w_{\varphi_i}^2 \sum_{j=1}^{np} [(\phi_{ij}(\alpha) - \phi_{ij})^{\text{analy}} - (\phi_{ij}^{\text{damaged}} - \phi_{ij}^{\text{undamaged}})^{\text{exp}}]^2,$$

where the number of measured points is 'np' and ϕ_{ij} is the jth component of ith mass in the normalised mode shape. This can also be assessed from the frequency changes combined with the respective mode shape. The following function is then valid:

$$
\begin{aligned}
J = \sum_{i=1}^{nm} \overline{w}_{\lambda i}^2 & \left[\left(\frac{\lambda_i(\alpha) - \lambda_i^0}{\lambda_i^0} \right)^{\text{analy}} \right. \\
& \left. - \left(\frac{\lambda_i^{\text{damaged}} - \lambda_i^{\text{undamaged}}}{\lambda_i^{\text{undamaged}}} \right)^{\text{exp}} \right]^2 \\
+ \sum_{i=1}^{nm} \overline{w}_{\varphi i}^2 & \sum_{j=1}^{np} [(\phi_{ij}(\alpha) - \phi_{ij})^{\text{analy}} \\
& - (\phi_{ij}^{\text{damaged}} - \phi_{ij}^{\text{undamaged}})^{\text{exp}}]^2 .
\end{aligned}
$$

2.11. Planning and Management

Damage detection or identification is very vital for health monitoring. There are four levels of damage detection, which are commonly practised in health monitoring.

- Level 1 deals with determination of presence of damage.
- Level 2 deals with the determination of location of damage.
- Level 3 deals with the quantification of severity of damage.
- Level 4 deals with the prediction of remaining service life of the structure.

There are many global methods to detect damage, as recommended by ISIS Canada Research group, which has developed guidelines for SHM. Before planning SHM and managing the SHM process, if vibration monitoring is carried out to estimate the damage in a given system, specific objectives are as follows (as per ISO:2002):

- evaluation of accuracy and constructability;
- evaluation of structural performance during construction and upon completion;
- assessment of safety of bridges during construction and upon completion;
- evaluation of serviceability upon completion of construction;

- determination of structural characteristics for updating the numerical model;
- feedback to update the structural design process.

2.11.1. *Structural assessment algorithm*

Maintenance of infrastructure depends upon several factors: (i) importance of the structure; (ii) cost of maintenance; and (iii) new demands on the structure due to additional loads, if any. Under normal conditions and load effects, degradation of material or accidents, the structure may lose its functionality. Therefore, two types of events can trigger an inspection of a structure:

(1) periodic inspection, which will be determined based on the maintenance strategy;
(2) invited inspection (I^2), which will be generally triggered by any external event.

The external event can be an alarm raised by the public on observing any damage, overloading of a bridge, observations of toll-booth operators and contractors of highways and permanent way inspectors in case of railway. Therefore, monitoring civil engineering infrastructure involves inspection where the main objective is to diagnose the structure, its support conditions and then to recommend (or advise) the decision makers in choosing one of the following options:

- Building a new structure in place of the old ones: This can be a result of severe damage that has occurred in the structure, which has reduced the performance level to a very low index.
- Restricted use of the existing structure in case there is less significant damage: For example, in case of a highway bridge, traffic loads, types of vehicles on the bridges and speed of the vehicles can be restricted.
- Strengthening the existing ones to increase or enhance the performance level of the structure.

- Recommending continuous monitoring in case the performance of the structure does not meet the desired safety requirements under extreme loads such as flood, earthquake, hurricanes, etc.
- Suggestion for development of an alarm system integrated with the health monitoring scheme so that the safety of the public is enhanced.

2.11.2. *Planning process*

In case of planning for SHM deployment, the following factors are important:

(a) *Establishment of objectives of SHM inspection*

This should focus on planning, analysis, operating and evaluating the complete data of the SHM inspection process. During the operation phase of the SHM scheme, one should also check a typical outcome of the scheme or process and its accuracy, so that benefits of SHM are not over estimated. Further, the objectives should be within the budget of the SHM system; costs should include implementation up to the control stage.

(b) *Establishment of a convenient budget for carrying out SHM*

The cost of SHM should include control strategies and buffer system, which is planned as a stand-by.

(c) *Choosing appropriate analysis tools*

Analysis tools should be chosen such that they are capable of identifying the parameters which are to be monitored or computed. Extraneous interpolation of data of the parameters should be avoided. If the structure is under static load, then there is no need to perform complicated dynamic analysis.

(d) *Design or development of a suitable sensor system layout*

Monitoring system should be carefully designed. It must include all vital parts of the health monitoring scheme. One has to decide the type of monitoring, namely, local monitoring or global monitoring, long-term or STM, periodic or triggered monitoring. This shall enable decision on the layout of the sensor system.

(e) *Design of topography of the sensor system, acquisition system and communication system*

This is vital in SHM planning. The sensor system designed and adopted should fulfil the strategies of monitoring. For example, if an LTM is suggested, then sensors should be robust and should be capable of giving reliable results over a long period of time. The commonly used approach is to glue the sensors to the surface of the structures, such as strain gauges, which is not advisable in case of LTM because sensors and glue will be affected by humidity, temperature, ultraviolet lighting, etc. it is also important to decide the sensor location, layout and scalability. Layout should be planned in such a manner that they are scalable and capable of reorienting themselves to a new set of command control. The whole layout should be well documented to execute it comfortably.

(f) *Design of acquisition and storage system*

Acquisition system actually controls the overall execution of the SHM process and hence should be carefully designed. All equipment required to collect the data, interpret and store the data in physical terms should be planned and well documented. One should also plan for sufficient back-up scheme for equipment and sensor so that in case of any breakdown, monitoring can be done continuously.

(g) *Design of communication systems*

They are also vital and should be planned to avoid loss of data in communication especially when the system triggers an alarm. In the latter case, the back-up system is necessary to maintain safety even when the system is collapsing.

(h) *Design of evaluation method of SHM*

Evaluation system is designed in such a manner that it remains compatible with the monitoring operation. It should also be easy to handle and install, and capable of tracking the pattern recognition of the behaviour. Pattern recognition is faster and effective to assess the condition of the structure.

2.12. Vibration-Based Monitoring

Civil engineering structures with the recent advancement of sensors, actuators and computational capabilities have become smart structures. They are intelligent enough to undergo a self-diagnosis to develop early warnings in case of any critical health state. SHM methods should therefore address certain basic themes or purposes as follows:

- SHM methods should deal with reliable functional components to avoid malfunctioning of the system or scheme and thus ensure public safety.
- They should be effective and efficient such that functional losses to the structural system can be avoided. If the structural system is not functioning properly, it will face a downtime for repair, which can lead to economic loss.
- They should enable revisiting the design principles towards light weight structures because maintenance and assessment are more effective in light weight structures.

Considering the factors such as causes for the damage, material and functional degradation, load path shifting, a damaged structure can be expressed in terms of a coupled system as follows:

$$M(\theta_d, \theta_e, x, t)\ddot{x} + g(x, \dot{x}, \theta_d, \theta_e, t) = f_{\mathrm{op}}(t) + f_{\mathrm{exp}}(t),$$

where M is the mass matrix, g is the force vector, which is a function of elastic damping force and it depends upon velocity, displacement and time, θ_d is the damage parameter arising from crack formation, loss of stiffness and (or) mass, θ_e indicates the influence of environmental forces and operational conditions on the structure's health. For example, temperature, humidity, change of mass distribution can be the variables. f_{op} is the operational load and f_{exp} is the experimental load, which are the scaled magnitude of operational loads.

Damage function θ_d is nonlinear and it can be expressed as follows:

$$\theta_d = \Gamma(\theta_d, \theta_e, x, \dot{x}, t).$$

While performing such analyses, evaluation of damage and dynamic response under damaged condition takes place in two different timescales. One may be the slowly varying component, while the other may be the rapidly varying component. Therefore, it is necessary that θ_d shall account for such variations carefully.

2.12.1. *Damage identification in linear systems*

Equation of motion of 'n' degree-of-freedom system can be expressed as follows:

$$m\ddot{x} + c\dot{x} + kx = f(t).$$

If the system is undamped (or lightly damped), then the characteristic features of the system, such as natural frequency and mode shape, can be determined using the classical eigensolver theory as follows:

$$\left(k - \omega^2 m\right)\phi_n = 0.$$

Alternatively, one can use correction parameters to represent the modal changes in the element level of the structural system. This can be expressed as follows:

$$\Delta k = \sum_j k_j \Delta_{aj},$$

$$\Delta c = \sum_j c_j \Delta_{aj},$$

$$\Delta m = \sum_j m_j \Delta_{aj},$$

where the general damage parameter expressed as θ_d is now replaced with a linear matrix correction Δ_{aj}. The correction parameter, which localises and quantifies the damage, can be determined by solving the inverse problem. This can be done by minimising the weighted sum of components of data error. Minimising the following function with ε, we get

$$J = \varepsilon^T w_\varepsilon \varepsilon + \Delta_a w_a \Delta_a,$$

$$\varepsilon = S\Delta_a - \nu,$$

where S is the sensitivity matrix, which can be computed from the first-order partial derivative of the dynamic quantities with respect to the correction parameter 'a'.

Weighted matrices are given by

$$J = \varepsilon^T w_\varepsilon \varepsilon + \Delta_a w_a \Delta_a.$$

Alternatively, one can also use the frequency-band method for damage detection called frequency-based damage detection (FBDD). A single damage index for jth member of any structural system is given by

$$\text{DI}_j = \left[\sum_{i=1}^{nm} e_{ij}^2 \right]^{-1/2},$$

where DI is the damage index at the jth element of the structure, nm is the number of modes considered for the analysis, e_{ij}^2 is the localisation error for ith mode in the jth element of the structural system and is given by

$$e_{ij} = \frac{Z_i}{\sum_{k=1}^{nm} Z_k} - \frac{F_{ij}}{\sum_{k=1}^{nm} F_{kj}},$$

$$Z_i = \frac{\delta\omega_i^2}{\omega_i^2},$$

where Z_i indicates the fractional change caused by the change in the ith eigenvalue

$$\delta\omega_i^2 = \omega_i^{*2} - \omega_i^2,$$

where ω_i^{*2} is the changed natural frequency and F_{ij} is the fraction of modal strain energy (MSE) for ith mode, stored in the jth element of the structure. This is given by

$$F_{ij} = \frac{\{\phi_i^T\}\,[K_j]\,\{\phi_i\}}{\{\phi_i^T\}\,[K]\,\{\phi_i\}},$$

where $\{\phi_i\}$ is the ith mode shape vector, K is the system stiffness matrix and K_j is the contribution of jth element to the system stiffness matrix. Once Z_i is determined experimentally, then F_{ij} can be determined numerically.

2.12.2. *Specific advantages of vibration-based SHM*

The basic feature of vibration-based monitoring is to identify the changes in structural characteristics, such as mass, stiffness and damping, caused by the presence of damage. Therefore, by examining the changes in the measured values of structural characteristics (vibration characteristics) and then solving the inverse problem, we shall quantify the unknown changes in the original system. Vibration-based SHM consists of five steps:

- Step 1 deals with the measurement of structural dynamic response in terms of acceleration and displacement. This is essentially done by set of sensors, acquisition system and transmission of data. Acceleration, velocity and displacement are key issues in the measurement. They will generate a big volume of data. Therefore, acquisition and transmission, which includes data storage also, should be carefully designed to avoid any loss of data.

- Step 2 deals with the characterisation of initial structural model through both static and dynamic tests. Initial characterisation provides the base line for comparing the response of the structure before and after the damage while the vibration characteristics of the functional structure are obtained by continuous monitoring, data acquired will be compared with the base line model.

- Step 3 involves continuous monitoring and damage localisation of the structure. During this process, data acquired and stored continuously should be analysed for its comparison with the baseline model. Any significant change in the vibration characteristics can lead to damage localisation.

- Step 4 deals with the detailed finite element analysis to update the structural model with the input from the observed damages. So, a constant update of the finite element model is required to be carried out.

- Step 5 is the evaluation of the structural performance of the updated model. At this stage, the updated model shall reveal the health status of the present structure.

These steps are given as a flow chart in Figure 2.4. The data measurement is carried out through data sensing, data transmission and data analysis. Based on this, one can apply it to real-time vibrating system, which gives the initial characterisation of the system. However, one can also perform parallel, continuous monitoring. Through initial characterisation, results can be obtained from either static test or dynamic test. Both of these data will be useful to prepare a baseline model. Similarly, from the continuous monitoring, one can achieve the vibration signature records, which can be used in performing modal analysis for structures in operation.

Modal analysis will be different from that of the conventional characterisation of the system based on which the baseline model has been prepared. Based on this value, one can subsequently achieve damage localisation and finally the model can be updated to carry out performance evaluation. The performance evaluation will

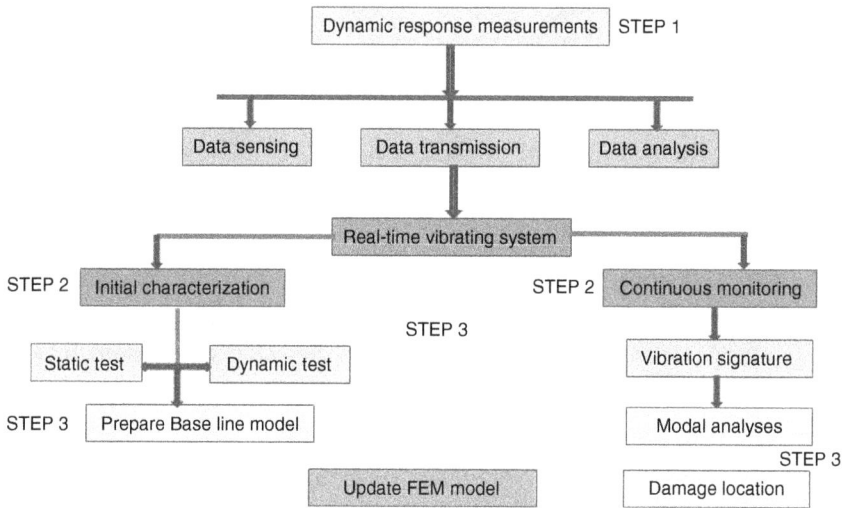

Figure 2.4. Vibration-based monitoring.

subsequently lead to two issues: (i) capacity building of the structural system; and (ii) service life prediction of the system.

2.13. SHM Methods

2.13.1. *Method using frequencies and mode shapes*

It is vital to understand the changes in structural characteristics. This can be readily identified by noticing the changes in natural frequencies. For example, change in stiffness and mass changes the eigenvalues can be modelled as follows:

$$\{Z\} = [F]\{\alpha\} + [G]\{\beta\},$$

where $\{Z\}$ is the vector of measured frequency changes, $\{\alpha\}$ and $\{\beta\}$ are the vectors of changes in the system related to stiffness and mass, respectively; $[F]$ and $[G]$ are called sensitivity matrices.

In order to compute $\{\alpha\}$ and $\{\beta\}$, which quantify the changes in stiffness and mass matrices, one needs to compute the sensitivity matrices. They can either be computed analytically from the eigenvalues or numerically utilising perturbation methods using the finite element model. However, research studies conducted earlier showed that the change in mass matrix, before and after the damage, is negligible. Therefore, the sensitivity equation can be reformulated as follows:

$$Z_i = \sum_{j=1}^{NE} F_{ij}\alpha_j,$$

where Z_i indicates the fractional change of ith eigenvalue (frequency), α_j indicates the fractional reduction in the ith stiffness parameter, F_{ij} is expressed as a fraction of MSE for the ith mode stored in jth element of the structure. F_{ij} can be expressed as follows:

$$F_{ij} = \frac{\{\phi_i^T\}[K_{ij}]\{\phi_i\}}{\{\phi_i^T\}[K]\{\phi_i\}},$$

where $[K]$ and $[K_{ij}]$ are global and elemental stiffness matrices, respectively. Once the stiffness matrix of the complete structural system and mode shape are known, then F_{ij} can be generated numerically. Subsequently, to obtain the relative damage, the sensitivity equation to obtain relative damage can be estimated from the following relationship:

$$\frac{Z_m}{Z_n} = \frac{\sum_{j=1}^{NE} F_{mj}\alpha_j}{\sum_{j=1}^{NE} F_{nj}\alpha_j}.$$

For example, if only one element is damaged, then the above equation reduces to

$$\frac{Z_m}{Z_n} = \frac{F_{mq}}{F_{nq}},$$

which is unique for the qth location. Then the error index is given by

$$e_{ij} = \frac{Z_m}{\sum_{k=1}^{NM} Z_k} - \frac{F_{mq}}{\sum_{k=1}^{NM} F_{kq}},$$

$e_{ij} = 0$, which indicates the damage at the jth location.

Damage sensitivity is given by the following relationship:

$$\frac{\delta\omega_i^2}{\omega_i^2} = \eta S_{ik} \left(\frac{a_k}{H}\right)_i^e,$$

where $\frac{\delta\omega_i^2}{\omega_i^2}$ is the fractional change in the eigenvalue in the ith mode, $\left(\frac{a_k}{H}\right)_i^e$ is the dimensionless crack size, which is normalised to the depth of the member (H), η is the shape factor accounting for the geometry of the mass, S_{ik} is the sensitivity of the kth location in the ith MSE. If there is a fractional change in the eigenvalue of the system measured experimentally, then one can easily determine the crack size from the above equation. But this equation has a limitation: if only one damage is present, this equation can be used to locate the damage. In case of multiple damages, or damage at multiple locations, it is not applicable. Based on the changes in stiffness and mass matrices, one can observe changes in eigenvalues, before and after the damage, to identify the damage location and crack size as explained before.

2.14. Shear Model Building

Let the undamped shear building is expressed by the equation of motion as follows:

$$m\ddot{x} + kx = f(t).$$

The characteristic equation to determine the eigenvalues and mode shape is given by

$$\left(k - \omega_i^2 m\right)\phi_i = 0,$$

where ω_i is the eigenvalue and ϕ_i is the corresponding eigenvector or mode shape. Stiffness matrix of the shear building is given by

$$[K] = \begin{bmatrix} k_1 + k_2 & -k_2 & \cdot & \cdot & & \cdot \\ -k_2 & k_2 + k_3 & \cdot & \cdot & & \cdot \\ \vdots & \vdots & \vdots & \vdots & \vdots & \vdots \\ \cdot & \cdot & \cdot & \cdot & k_{n-1} + k_n & -k_n \\ \cdot & \cdot & \cdot & \cdot & -k_n & k_n \end{bmatrix}.$$

Mass matrix is given by

$$[M] = \begin{bmatrix} m_1 & 0 & \cdot & \cdot & & \cdot & \\ 0 & m_2 & \cdot & \cdot & & \cdot & \\ \vdots & \vdots & \vdots & \vdots & \vdots & & \vdots \\ \cdot & & \cdot & \cdot & m_{n-1} & 0 \\ \cdot & & \cdot & \cdot & 0 & m_n \end{bmatrix}.$$

Expanding the above equations for l and r modes and rearranging terms of stiffness and mass matrices, the following equation is generated:

$$\begin{bmatrix} \phi_{l1} & -\omega_l\phi_{l1} & \phi_{l1} - \phi_{l2} & \cdot & & \cdot & \\ \phi_{r1} & -\omega_r\phi_{r1} & \phi_{r1} - \phi_{r2} & \cdot & & \cdot & \\ \vdots & \vdots & \vdots & \vdots & \vdots & & \vdots \\ \cdot & \cdot & & \cdot & \phi_{ln} - \phi_{ln-1} & \omega_l\phi_{ln} \\ \cdot & \cdot & & \cdot & \phi_{rn} - \phi_{rn-1} & \omega_r\phi_{rn} \end{bmatrix} \begin{Bmatrix} k_1 \\ m_1 \\ \vdots \\ k_n \\ m_n \end{Bmatrix} = 0,$$

or simply

$$[B]\{b\} = 0.$$

If $m_n = 1$, then the above equation will reduce to the following form:

$$[B']\{b'\} = \begin{Bmatrix} 0 \\ 0 \\ \vdots \\ \omega_l\phi_{ln} \\ \omega_r\phi_{rn} \end{Bmatrix},$$

where $[B']$ is of the order $(2n - 2) \times (2n - 2)$ in which the last two rows and columns of $[B]$ are eliminated, $\{b'\}$ is of the order $(2n-2)\times 1$ vector in which last two rows of $\{b\}$ are eliminated. Now, solving for the unknown mass and stiffness parameters, we get

$$\{b'\} = \left[[B']^T[B']^{-1}\right][B']^T \begin{Bmatrix} 0 \\ 0 \\ \vdots \\ \omega_l \phi_{ln} \\ \omega_r \phi_{rn} \end{Bmatrix},$$

$$k_n = \frac{\omega_l \phi_{ln}}{\phi_{ln} - \phi_{ln-1}}.$$

Now, the mass and stiffness parameters obtained from the above equation are relative values of m_n because it is considered to be unity. This method has salient advantages as follows:

(1) Only mode shape and frequency of two modes are required.
(2) This can be applicable only to shear model buildings.
(3) This is valid only for undamped systems.

2.15. Damage Identification Using Lumped Mass Model

The change in eigenvalues can enable damage detection. Let us consider the dynamic equilibrium of forces acting at the jth storey by considering all the storeys above. The following relationship holds good:

$$k_{i(j)}d_{i(j)}(t) + C_{i(j)}\dot{d}_{i(j)}(t) = f_{i(j)}(t),$$

where $k_{i(j)}$ and $C_{i(j)}$ are stiffness and damping parameters of the jth storey, respectively, $f_{i(j)}(t)$ is the inertia force acting on the storey mass on jth storey and all the storeys above it. Then, the following relationship holds good:

$$f_{i(j)}(t) = -\sum_{k=j}^{N} m_k \ddot{x}_{i(k)}(t) = -\omega_i^2 e^{\omega_i^t} \sum_{k=j}^{N} m_k \phi_{i(k)},$$

where N is the number of storeys, m_k is the mass of the kth storey, $d_{i(j)}(t)$ is the relative displacement between jth and $(j-1)$th storeys.

For ith mode of vibration, this can be expressed as

$$d_{i(j)}(t) = X_{i(j)}(t) - X_{i(j-1)}(t) = \left\{\phi_{i(j)} - \phi_{i(j-1)}\right\} e^{\omega_i t},$$
$$\dot{d}_{i(j)}(t) = \dot{X}_{i(j)}(t) - \dot{X}_{i(j-1)}(t) = \left\{\phi_{i(j)} - \phi_{i(j-1)}\right\} e^{\omega_i t}.$$

Once, the eigenvector and eigenvalue are measured for ith storey, then the equation for jth storey at each time interval using short-data length can be solved. Short-data length is approximately equal to the natural frequency of the building. One can repeat this procedure for other storeys to determine the stiffness and mass parameters of each storey. Some salient points of this method are as follows: (i) it assumes a known storey mass; and (ii) it is applicable for lumped mass, shear building model.

2.16.   Damage Identification Using Element Modal Stiffness

In a linear undamped structure, ith modal stiffness is given by

$$K_i = \phi_i^T [K] \phi_i,$$

where $[K]$ is the complete stiffness matrix of the entire structure. Now, combination of jth member to ith modal stiffness is given by

$$K_{ij} = \phi_i^T [k_j] \phi_i,$$

where k_j is the jth member contribution to the total stiffness matrix $[K]$. Now, fraction of modal energy of ith mode contributed by jth member is called modal sensitivity. Modal sensitivity is given by the following relationship:

$$F_{ij} = \frac{K_{ij}}{K_i}.$$

The above equation is valid for undamaged structure. This is modified for a damaged structure. The modal sensitivity for a damaged structure is given by the following set of equations:

$$F_{ij}^* = \frac{K_{ij}^*}{K_i^*},$$

$$K_{ij}^* = \phi_i^{*T}[K_j^*]\phi_i^*,$$

$$K_i^* = \phi_i^{*T}[K^*]\phi_i^*,$$

$$K_j = E_j[K_{j0}],$$

$$K_j^* = E_j^*[K_{j0}],$$

where E_j and E_j^* represent material stiffness property related to undamaged and damaged states of the structure, respectively. $[K_{j0}]$ is assumed to be unchanged even after the damage. The basic assumption is that the modal sensitivity for ith mode and jth member remains unchanged before and after the damage. Mathematically, this can be expressed as

$$\frac{F_{ij}}{F_{ij}^*} = \frac{K_{ij}^* K_i}{K_i^* K_{ij}} = 1.$$

Now, the damage index for jth member is defined as follows:

$$\beta_j = \frac{E_j}{E_j^*}.$$

Substituting from the earlier equations, we get

$$\beta_j = \frac{\nu_{ij}^* k_i}{\nu_{ij} k_i^*},$$

$$\beta_j = \frac{\phi_i^{*T}[K_{j0}]\phi_i^* K_i}{\phi_i^T[K_{j0}]\phi_i K_i^*}.$$

The damage index can also be approximated as

$$\beta_j \cong \left[\frac{\phi_i^{*T}[K_{j0}]\phi_i^* + \sum_{k=1}^{NE}\phi_i^{*T}[K_{j0}]\phi_i^*}{\phi_i^T[K_{j0}]\phi_i K_i^* + \sum_{k=1}^{NE}\phi_i^T[K_{j0}]\phi_i K_i^*}\right]\left(\frac{K_i}{K_i^*}\right).$$

Now, the normalised damage indicator is given by

$$Z_j = \frac{\beta_j - \bar{\beta}}{\sigma_\beta},$$

where $\bar{\beta}$ is the mean value and σ_β is the standard deviation. Severity of damage can be estimated as

$$E_j^* = E_j \left(1 + \frac{dE_j}{E_j} \right) = E_j(1 + \alpha_j),$$

where $\alpha_j = \frac{\gamma_{ij} k_i^*}{\gamma_{ij}^* k_i} - 1.$

2.17.   Damage Identification Using Modal Strain Energy

This is a two-stage process. In the first stage, locate the damage using change in MSE of the element. In the next stage, determine the extent of damage by the iterative scheme. MSE of jth element of both undamaged and damaged case is given by

$$\text{MSE}_{ii} = \phi_i^T [k_j] \phi_i,$$
$$\text{MSE}_{ij}^d = \phi_{di}^T [k_{dj}] \phi_{di},$$

where 'd' indicates the damaged state, k_j is the element stiffness matrix of jth element, ϕ_i is the ith mode shape. Change in MSE ratio is indicated as MSRCR and is given by the following relationship:

$$\text{MSECR}_j^i = \frac{\left| \text{MSE}_{ij}^d - \text{MSE}_{ij} \right|}{\text{MSE}_{ij}}.$$

The above equation is a meaningful indicator of damage elements. Damage elements will have a significant change in the stiffness as stiffness will be degraded. Therefore, change in stiffness for the damaged element can be expressed as a fractional change of the elemental stiffness matrix. The stiffness of the damaged member is given by

$$K^d = K + \sum_{j=1}^{L} \Delta k_{ij} = K + \sum_{j=1}^{L} \alpha_J k_{ij} \quad \text{for } -1 < \alpha_j \leq 0.$$

Now, change in MSE is expressed as follows:

$$\text{MSEC}_{ij} = 2\Delta \phi_i^T k_j \phi_i + \alpha_j \phi_i^T k_j \phi_i.$$

In the above equation, α_j is unknown. To start with, this is assumed to be zero and iteration is set in. Thus, the modal strain energy correction (MSEC) is given by

$$\text{MSEC}_{ij} = 2\Delta\phi_i^T k_j \phi_i.$$

Now, the MSE change can be determined for both damaged and undamaged states by using the appropriate stiffness and mode shape. In the undamped case, the following equation holds good:

$$[(k + \Delta k) - (\omega_i + \Delta\omega_i) M][\phi_i + \Delta\phi_i] = 0.$$

Now, $\Delta\phi_i$ is expressed as a linear combination of mode shape of undamaged system and it is given by

$$\Delta\phi_i = \sum_{k=1}^{n} d_{ik}\phi_k.$$

By substituting the above equation and neglecting higher order terms, we get

$$d_{ir} = -\frac{\phi_r^T \Delta k \phi_i}{(\omega_r - \omega_i)} \quad \text{for } r \neq i.$$

MSE change is then given by

$$\text{MSEC}_{ij} = 2\phi_i^T k_j \left(\sum_{r=1}^{n} -\frac{\phi_r^T \Delta k \phi_i}{(\omega_r - \omega_i)} \phi_r \right).$$

The earlier equations of MSEC can be simplified as

$$\text{MSEC}_{ij} = \sum_{j=1}^{L} 2\alpha_p \phi_i^T k_j \left(\sum_{r=1}^{n} -\frac{\phi_r^T \Delta k \phi_i}{(\omega_r - \omega_i)} \phi_r \right) \quad \text{for } r \neq i.$$

Once the damage is located, then the damage sensitivity can be determined as follows:

$$\begin{Bmatrix} \text{MSEC}_{i1} \\ \text{MSEC}_{i2} \\ \vdots \\ \text{MSEC}_{ij} \end{Bmatrix} = \begin{bmatrix} \beta_{11} & \beta_{12} & \cdot & \cdot & \beta_{1p} \\ \beta_{21} & \cdot & \cdot & \cdot & \cdot \\ \vdots & \vdots & \vdots & \vdots & \vdots \\ \beta_{j1} & \cdot & \cdot & \cdot & \beta_{jp} \end{bmatrix} \begin{Bmatrix} \alpha_1 \\ \alpha_2 \\ \vdots \\ \alpha_p \end{Bmatrix},$$

where p is the number of suspected damage sites or locations, j is the number of elements considered to compute the MSEC

$$\beta_{st} = -2 \sum_{r=1}^{n} \phi_i^T k_s \frac{\phi_r^T k_i \phi_i}{(\omega_r - \omega_s)} \phi_r \quad \text{for } r \neq i,$$

where β_{st} is the element sensitivity coefficient of MSEC. MSEC is experimentally measured and then substituted in the above equation to obtain the fractional change in the stiffness of the damaged system or elements. Once, the initial value of α_p is obtained, values of MSEC can be updated for each iteration until convergence is reached.

2.18. Damage Identification Using Visual Inspection

Damage identification can also be done using visual inspection methods. One of the serious limitations of this method is that it affects the decision-making process and resource utilisation significantly. The issues in damage identification using visual inspection are as follows:

- *Timing interval*: The inspection frequency of visual inspection methods can be selected as per the requirements of the structure, environmental conditions and operational loads. It is important to note that the static assessment of the structure may not be sufficient enough to identify damages which are critical. Therefore, continuous monitoring is preferred, for example, crack propagation which cannot be captured by visual inspection tools.
- *Interpretation of results of visual inspection method*: Visual inspection method strongly depends on the visual inspectors, their expertise and domain knowledge, their experience and training, etc. The result of visual inspection is a subjective assignment which may be inadequate to compare with true assessment. This is due to the fact that visual inspection team may not be experienced, visual inspection guidelines used by different agencies may differ. There are no set standard guidelines for visual inspection.
- *Accessibility*: It is very important to know that the effective results of visual inspection depend on the physical accessibility of

Table 2.4. Sensors used for measuring various parameters.

Sensor type	Functionality	Purpose
Accelerometer	Vibrations	Modal analysis
Strain gauge	Surface strain	Stress–strain responses
Anemometer	Wind velocity and direction	Wind load assessment
Inclinometer	Inclination	Pier settlement in bridges
GPS receivers	Displacement and response	Model validation
Sonar	Pier tip elevation of bridge	Scour detection
Reference electrodes	Voltage potential of steel	Corrosion monitoring

the visual inspection team to the surface of prospective damage. If sufficient accessibility is not provided, internal irregularities cannot be interpreted from the results or reports of visual inspection method. This is a very serious limitation (Table 2.4).

2.19. SHM Challenges in Comparison to Alternate Methods for Health Monitoring

- System complexity: It is not the complexity of the structural system, rather it is complexity of the SHM system. This is dependent on the size and complexity of the structure being monitored. It also depends on the functional characteristics of the structural system, for example, an automated multi-functional SHM system which is integrated with an alert monitoring system. Such systems require a complex and robust network software which is highly complicated. The SHM system complexity also depends on the expected remaining service life of the structure.
- SHM system maintenance: SHM system requires complicated network of sensors which are laid and controlled by complex hardware and software. There is a major problem of breakdown of the system itself. Therefore, it requires regular maintenance to sustain LTM. SHM systems need rigorous and continuous maintenance which can be of high expense. There are some tips to reduce the SHM system maintenance cost:

  ○ Reduce the system redundancy of the structure.

○ Avoid total breakdown and provide renewable power source to the hardware of SHM systems. This eliminates the need to change battery in case of wireless sensors.

○ One needs to employ adequate IT professionals to ensure ongoing functional condition of the SHM system.

- Automated data analysis: Let us consider an SHM system equipped with automated control triggering and automated communication in terms of alert monitoring. It is very important to know that all data collected by the sensors may not be relevant to identify the potential damage. Therefore, it is important that the data analysis capability should be enhanced in case of automated data analysis. Further, the SHM system should be well trained for the set of data to which it needs to respond automatically. Sometimes, it may cause false alarming also.

- Liability and responsibility: It is very important that in case of continuous monitoring, data are acquired from the structural system on a continuous basis. In such a situation, all data collected need not be processed. Therefore, the authenticity of processing the valid data is a challenge because the data should be reliable, it should be taken from the required source and not been interpolated. For example, if any data related to failure or collapse of the structure is missed by any chance, then who holds the responsibility? This throws up major accountability issues.

2.19.1. *Comparison of visual inspection and SHM methods*

(1) **Functionality:** In real sense, the overall functionalities of full-scale SHM and visual inspection are not exclusively different. But, one important difference is the frequency or the interval at which visual inspection is carried out. Visual inspection methods have discrete and infrequent time intervals, whereas SHM methods have pre-set time intervals or are continuously monitored. Therefore, SHM methods have the potential to generate information even on a daily basis. Visual inspection methods cannot do this. But, one major functional advantage of

the visual inspection method is that the scope of visual inspection method is not only limited to damage detection, but it also leads to broad evaluation of the complete structure (preliminary assessment of the structure is possible with visual inspection). Even completely automated SHM methods cannot execute or extend the damage detection scenario to the complete structure.

(2) **Cost:** In both the cases, the cost essentially depends on the characteristics of the structure monitored. However, cost implementation of the SHM system will be very high if it is not required for the structural system, which means that SHM methods are applicable generally to structures of high importance. Visual inspection methods can be applied to all types of structures. Cost of visual inspection is subjected to the extent of details required from the visual inspection. It also depends on the inspection frequency. There are three factors based on which cost can be compared (Table 2.5):

- upfront cost;
- maintenance or operation cost;
- return on investment.

Table 2.5. Comparison of SHM and visual inspection methods.

SHM methods	Visual inspection methods
Majority of upfront cost in DHM methods is towards hardware or software components	Major cost is towards technical expertise of labour and use of advanced equipment to conduct visual inspection
Maintenance cost depends on the longevity and health of the structure. The cost is essentially towards data acquisition and data management	As such, there is no special operational cost involved, except in case of inaccessible locations like offshore structures where the labour will need to be insured
The return on investment is slow. This is due to the fact that the effectiveness of SHM will be realised only when the maintenance cost goes down in comparison to the rebuilding cost	The return on investment is quick and visual. This is due to the fact that there is an immediate perception of visual inspection results based on which maintenance or repair is initiated

(3) ***User resistance***: It is very important to note that SHM methods deploy advanced features of IT industry. Advanced sensors, acquisition systems, communication systems, data management, etc. have become an inherent part of SHM methods. There will be a significant shift towards IT-based system maintenance from conventional civil engineering maintenance. This involves resistance from the user in terms of using updated software, employing more IT professionals, etc. Visual inspection is done periodically or at scheduled intervals. These periods of inspection intervals are well planned and become a part of the technical maintenance of the structural system. More or less, visual inspection is contractual, which means it does not demand any additional training or knowledge towards civionics for the user. Visual inspection methods are carried out by expert third parties, but in general, user resistance is vital and important to successfully implement and execute SHM on the structures. It is also important to note that both visual inspection and SHM methods are to be combined for effective results. Depending solely on either one of them is not a good practice. Combined approach will be more successful for early detection of structural problems and reduces human error. Even though initial investment towards wireless sensors of SHM method may be expensive in comparison to visual inspection method, the added functionality and timeliness of decision support add more value towards higher initial investment.

2.20.   Comparison of Vibration-Based SHM Methods

The fundamental idea of vibration-based monitoring is to detect damage on the fact that damage-induced vibration changes structural properties, such as mass, stiffness and damping. Detecting these changes in comparison to that of an undamaged model is useful to detect damage. The parameters used for comparison are frequency, mode shape and modal damping. For example, reduction in stiffness intuits the formation of cracks. Therefore, damage can be identified by change in stiffness characteristics of the structure.

2.20.1. *Natural frequency-based methods*

In general, these methods use natural frequency as the basic feature for damage identification. Identification of damage is different from the location of damage. These methods are a good choice for a simple reason, natural frequencies of a structural system can be readily measured at few accessible points on the structure. They are less contaminated by other noise data which make it more powerful.

2.20.1.1. *Multiple damage location assurance criterion*

This is a statistical correlation between analytical prediction of change in frequency and the measured frequency. Multiple damage location assurance criterion (MDLAC) is actually a function of damage extent vector (δ_D). It is given by

$$\delta_D = \frac{|\{\Delta f\}^T \delta_f \{\delta_D\}|^2}{(\{\Delta f\}^T \{\Delta_f\}) \cdot (\{\delta_f \{\delta_D\}^T\} \{\delta_f \{\{\Delta f\}^T \delta_f \{\delta_D\}\}\})},$$

where δ_f is the analytical prediction of frequency change and Δ_f is the measured frequency change. MDLAC provides a good prediction of both location of damage and size of damage (extent of damage) at one or even more sites.

2.20.1.2. *Single damage indicator*

This method is useful to locate and quantify the damage in flexural members. This method is good to locate and quantify the cracks in beams. This method uses change in natural frequency to detect the damage. Fractional change in modal energy is related to the fractional change in frequency which has occurred due to damage. Single damage indicator (SDI) is used to indicate damage location.

$$\text{SDI}_j = \left[\sum_{i=1}^{NM} e_{ij}^2 \right]^{-1/2},$$

where e_{ij} is the error index. This is used to represent localisation error for ith mode in jth location.

$$e_{ij} = \frac{Z_i}{\sum_{k=1}^{NM} Z_k} - \frac{F_{ij}}{\sum_{k=1}^{NM} F_{kj}},$$

where Z_i is the fractional change in ith eigenvalue due to damage which is given by

$$Z_i = \frac{\delta \omega_i^2}{\omega_i^2}.$$

Further, the sensitivity of the ith modal stiffness of jth element is given by

$$F_{ij} = \frac{K_{ij}}{K_i}.$$

2.20.1.3. *Spectral centre correction method*

This method is useful to detect damage based on auxiliary mass spatial probing. Spectral centre correction method (SCCM) correlates highly accurate natural frequency value based on the auxiliary mass location to detect damage. One important limitation of this method is that it is very difficult to compute natural frequency with high accuracy. Therefore, applying correction based on its correlation to auxiliary mass is difficult and complex.

Limitations of frequency-based methods are as follows:

- Most of the frequency-based methods are model dependant.
- Damage identification strongly depends on the Euler–Bernoulli beam theory. In this beam theory, crack formation is modelled as rotational spring. Euler–Bernoulli beam theory over-predicts natural frequency in short beams and also high frequency bending moments. Modelling crack as rotational spring is unsuitable for higher modes of vibration. It is also not suitable for deep or wide cracks. Therefore, frequency-based methods are more suitable for slender structures in particular.

Limitations related to frequency changes are as follows:

- Frequency changes caused by the presence of damage are lesser in comparison to those caused by other factors like environmental and operational conditions. For example, different studies show that frequency changes caused by environmental and operational conditions are usually in the range of 5–10%. Therefore, frequency changes caused by damage should be in this scale to make them noticeable, which means at least 5% change should be invoked in frequency changes, so that they can be recognised. The most difficult task is resulting in 5% change, which is possible only when the damage is severe or deep in nature. Frequency-based methods can be used to detect damage only when the damage is significant.
- Damage location proposed by these methods are generally ill-conditioned. It is seen that the damage with same severity, occurring in symmetric locations, will result in identical frequency changes. Damages with different severities occurring in different locations, which are asymmetric, can also cause identical frequency changes; this has been verified in few cases of measurements of natural frequency. Therefore, frequency-based methods of damage detection are not effective to detect or locate multiple cracks.

2.21. Methods Based on Mode Shape

Due to the following facts, mode shapes are preferred to locate damages in single and multiple locations:

- These methods use mode shapes and their derivatives for damage detection because mode shapes actually depict relative position of mass when the structural system is vibrating at a specific frequency. Therefore, they are mass sensitive (any change in mass will be reflected in the mode shape). This characteristic is used for damage detection.
- Mode shapes are very illustrative and interesting to use in case of multiple damage detection. It is due to the fact that they are highly sensitive to the presence of multiple damages.

- Mode shapes are less sensitive to environmental effects, such as temperature variation, in comparison to the variations or sensitivities of the corresponding natural frequencies.

Following are a few limitations of this method:

(1) To measure a mode shape, series of sensors are required to be placed in the lumped mass points (one at each mass point), which makes them expensive.
(2) Mode shapes are more prone to contamination by the presence of noise generated by machine vibrations, electrical appliances, etc. There is a possibility that they may also create a false damage location.

Change in mode shape is a common phenomenon used to detect damage. Change in mode shape between undamaged and damaged members can be used to detect damage. Mode shapes can be obtained either experimentally or numerically. Mode shapes are sensitive to damages occurring in critical areas like the mid-span of a simply supported beam. For accurate localisation of the damage, one generally uses signal pattern recognition, which requires additional processing. Since more sensors are required with higher sensitivity, this application in large structures is highly limited. Alternatively, people use mode shape analysis with modern signal processing.

2.22. Mode Shape Analysis with Signal Processing

Since change in mode shape between an undamaged member and damaged member is required for damage detection, it is important that mode shape in undamaged sections should be evaluated with higher accuracy. Numerical models with higher accuracy are also computationally expensive. Therefore, mode shape analysis can be carried out using signal processing. In such cases, there is no need for the numerical model. Mode shapes, which are estimated from the experimental investigations, can be used to detect damage; no detailed numerical model is essential.

The basic assumption in this method is that mode shape data of an undamaged structural system contain only low frequency signal in the spatial domain. Therefore, the presence of high frequency signal is an indicator of presence of damage in the structure. Thus, high frequency signals should be filtered out from the mode shape data, which is generally done using modern signal processing. There are two methods by which this can be processed — fractal dimension (FD) method and wavelet transform (WT) method. Both the methods cannot be used for damage quantification; they can be used only for damage detection.

(i) *FD method*

FD of a mode shape curve is given by

$$\text{FD} = \frac{\log_{10} n}{\log_{10} (d/L) + \log_{10} n},$$

where n represents the number of steps in the mode shape curve, d represents the distance between first point of sequence (P_1) and ith point of sequence (P_i), which actually provide the farthest distance, L represents total length of the curve which is given by

$$L = \sum_{i=1}^{n-1} d\,(P_i, P_{i+1}),$$

where d is the maximum distance between P and P_i. Peak of the FD curve can locate the damage and also the size of the damage. This can be inferred by the showing up of local irregularities of the fundamental mode shape, which is generally introduced or caused by the presence of damage. If one needs to include higher modes in the analysis, then the fractal dimension method is replaced by generalised fractal method (GFM) which is an improvement by a scale factor 's'. GFM for damage detection is given by

$$\text{FD} = \frac{\log_{10} n}{\log_{10} (d_s/L_s) + \log_{10} n},$$

where the subscript 's' indicates that the values are improved by scale factor.

$$d_s = \max_{1 \le j \le m} \sqrt{(y_{i+j} - y_i)^2 + s^2 (x_{i+j} - x_i)^2},$$

$$L_s = \sum_{j=1}^{M} \sqrt{(y_{i+j} - y_i)^2 + s^2 (x_{i+j} - x_i)^2}.$$

(ii) *WT method*

This method is used for advanced signal processing. It closely examines the signal of the mode shape with a multiple scale. The intention is to provide more details and approximations about the mode shape curve itself.

2.23. Mode Shape Curvature Usage in Damage Detection

Mode shape curvature (MSC) is the second derivative of the mode shape, which indicates high sensitivity to the presence of damage. Curvature of the mode shape can be approximated using a central difference technique as follows:

$$k_i = \frac{(\omega_{i+1} + \omega_{i-1} - 2\omega_i)}{h^2},$$

where ω is the modal displacement point, h is the spacing of the sensor used to obtain the mode shape. Change in the curvature of the mode shape is a good indicator of damage. It is useful to identify both the presence of damage and also the location of damage. Unfortunately in higher modes, modal shape curvature shows several peaks. This is a false indicator of damage. Curvature of lower mode shapes (fundamental mode shape) is very useful in damage identification.

2.24. MSE Method

It is important to note that a fractional change in MSE is also a good indicator of damage detection. It is an agreed fact that at least for bending elements (beams and plates), MSE can be directly related to the mode shape curvature.

All methods are effective only for small sensor spacing conditions. These methods cannot work if the sensors are widespread because interpolating responses between the sensor data will not enhance the accuracy of damage detection. All these methods are capable of locating single damage except MSC which can locate multiple damages. Fundamental mode shapes alone may not be always the most effective mode for damage detection. Even higher modes may become sensitive to damages. Natural frequency ratio between the damaged member and undamaged member can be a good indicator for the presence of damage. But this is true only in certain modes. MSC method is more robust in case measurements are disturbed by the presence of external noise created by machine vibrations and electrical signals.

2.25. Statistical Pattern Recognition

Vibration-based damage identification is useful in interpreting the local damage. But unfortunately due to the high complexities of the real structure under working conditions, it is very difficult to input the accurate mode shapes and frequencies to vibration-based methods for damage detection (Table 2.6). This may result in a large human error in damage identification. But there is a remedy for this. The remedy is that if the analysis is supported by semi-analytical methods, such as statistical pattern recognition (SPR), then a better accuracy can be seen.

Pattern recognition technique constitute sensors, which are used to measure strain and vibration of a structural member that produces signals. These signals are very sensitive to any change in strain or vibrations measured. They respond sensitively to environmental changes and operational conditions. One should group these changes associated with environmental conditions into a separate group of data. Changes in measurement, such as strain and frequency, are grouped as a pair to those environmental changes and operating conditions. Thus, a database is created, which is actually a group. Each

Table 2.6. Comparison of vibration-based methods.

Algorithm	Type	Parameters used	Basic assumption	Damage indicator
SDI	Model-based	Natural frequency should be measured and compared between a damaged and undamaged system	Single damage	This method works well at the element level
GFD	Response-based: this considers only the response of the damage state	The requirement is mode shape of damaged state and supplied as input with natural frequency	—	DI is confined to the sensor location
MSC	Response-based	This method demands measurement of mode shape curvature of both damaged and undamaged members	—	DI is confined to the sensor location

group will have a unique pair combination which is related to change in strain and the corresponding operating conditions. This is called as a pattern. A pattern that essentially arises from the signature of the measured signals is identified and this is now compared with the new pattern which is being recorded. Once the recorded signal change in their pattern matches with the existing patterns of database, then these changes are mapped to the corresponding damage locations. Therefore, pattern recognition is a machine learning process with the ability of the computer to identify and classify whether the observed data match or belong to a specific pattern (that is already existing) in the database. This can now expedite the decision-making process. This feature is very useful in case of automated SHM processes. Recognition patterns are of two types as follows:

(1) Supervised learning where the input patterns of vibration are compared with pre-defined groups in the database.

(2) Unsupervised learning where the pattern is compared with an undefined group, which may possibly become a new group in the database.

SPR method is a good simplification in SHM. SHM deals with experimental data which anyway have lots of uncertainties. These uncertainties can be handled in statistical models. Different algorithms are used to analyse the distribution of extracted features (strain, frequencies) to determine the damage location. However, these algorithms depend on the availability of data of damaged states. If the data are not available for the damaged states to compare, then one can use outlier analysis. One common application of outlier analysis recommended for SHM is the X-bar control chart.

2.26. Statistical Model Development

Following steps are important in damage identification by statistical model development (SMD):

Step 1: Measure a typical set of data, namely, steady state strain, live load strain, acceleration under live load, temperature effects on strain.

Step 2: When the structure undergoes damage, mean or variance of the extracted features changes significantly (accordingly).

Step 3: Auto-regression analysis will be carried out on the measured data preferably; first few auto-correlated regression coefficients say up to three are considered to obtain damage indications.

2.27. Damage Identification by Pattern Recognition

There is a prerequisite to apply this method. Several sample data on the damaged model, such as strain, acceleration, should be measured and pattern or group should be formed. The first set of data on the damaged section can be considered as the reference data, the rest can be considered as the test data. When new data are collected, they are compared to match the pattern with the reference data. So, SPR is useful to expedite the decision-making process in SHM.

2.28. Long-Term SHM

Long-term SHM refers to monitoring a structure over a large period of time varying from 10–15 years. It is essentially a static process. Fibre optic sensors, GPS receivers and corrosion sensors are most commonly used for LTM. LTM is a seven stage process.

Stage 1: Identify the structures that need LTM

The following structures need LTM:

(1) new structures including those have innovative aspects in design, construction, material, etc.;
(2) new structures associated with unusual level of risk, for example, geotechnical conditions of soil, seismic risk conditions (near fault), and aggressive corrosive environment;
(3) structures of strategic importance, such as offshore structures, nuclear reactors, etc.;
(4) existing structures whose disruption will affect the critical network, for example, important railway and highway bridges, reservoirs, etc.;
(5) new structures in which their features represent a large unit of subset of structures;
(6) existing structures with known deficiencies;
(7) existing structures which are recommended for rebuilding.

Stage 2: Risk analysis

Once the type of structure for LTM is decided, then one has to carry out risk analysis on these identified structures to list out possible events and degradations that can affect the structure, for example, corrosion, loss of pre-stress in highway girders, presence of creep, settlement of foundations, earthquake strike, impact of load effects and poor execution of public structures.

Stage 3: Response measurement

For the type of structures identified, for each risk identified in the previous step, one has to identify the corresponding consequence. For example, if corrosion is identified as risk in coastal structure which needs long-term health monitoring, then the expected consequence could be a chemical change and loss or degradation of section in terms of strength and durability. At this stage, based on the identified risk, approximate response of the identified structure is to be carried out. For example, if corrosion is risk-associated, then corrosion sensors should be chosen. If foundation settlement is risk-associated, then inclinometers should be chosen as the sensor type. At this stage, it is very important to choose the sensor requirements to measure the expected consequence. The desired output at this stage is list of quantities that need to be measured (monitored) along with their likely magnitude and location.

Stage 4: Design of SHM and sensor layout

If an inexperienced engineer carries out health monitoring, he will start from this stage. But, the previous three stages are important to make the health monitoring system more efficient and successful. The objective at this stage is to select the appropriate type of sensor, specifications of sensor (operational range), installation requirement (embedded or surface mounted), technical constraints of the sensors and budget constraints. It is always a good practice to include sensors based on different technologies. Do not choose the sensor of the same type. It is always better to have a mixture. Further, too many data acquisitions will become costlier and complex. Therefore, a simple design or layout is advised.

Different sensors can be connected to the same data logger. Several data loggers can be connected to the same data management system. It is very important to note that the data management system should integrate all data loggers to translate incoming data into a single format which can be forwarded. Most importantly, almost all vendors of sensors and data acquisition system provide their own software for data management. Therefore, it is very important to have a single integrated interface. A detailed design document should be prepared at this stage containing the following:

(1) list of sensors and type of sensors;
(2) layout architecture;

(3) installation plan and cable layout in case of wired sensors;
(4) installation procedure for every type of sensor;
(5) budget details.

Stage 5: Installation and calibration

One should follow the manufacturer's instruction to install sensors. Once they are installed and interconnected, then they need to be tested and calibrated. This is called site acceptance test (SAT). One needs to also fix the threshold values in case of automated alarming systems. Therefore, check this layout for its successful working. At this stage, a complete manual and calibration report should be generated.

Stage 6: Data acquisition and management

Data should be acquired and stored in the database. It should have an appropriate back-up and access authorisation which need to be checked. The major outcome of this stage is a complete document of the project management of data acquisition and management with log of events.

Stage 7: Data assessment

At this stage, the engineer should be able to identify a foreseen risk and expected outcome of the risk. A set of procedures to respond to any type of damage is to be created. For example, if the outcome is a simple degradation, then the procedure could be maintenance. If the outcome of recorded data is capacity reduction, then closer of the services should be recommended.

2.29. Non-Destructive Examination

There are three major areas where non-destructive examination is successful in the health monitoring process: (i) non-destructive testing (NDT); (ii) non-destructive evaluation (NDE); and (iii) non-destructive inspection (NDI). All these are very vital for the SHM scheme and become part of inspection methods and continuous or intermittent health monitoring. They are essentially used to detect the local failure of the structure. The most commonly practiced technique is ultrasonic inspection.

2.29.1. *Ultrasonic method*

In an infinite solid medium, elastic waves can propagate in two modes — pressure waves (P waves) and shear waves (S waves). If the medium is bounded, then these waves reflect at the boundaries to form a complicated wave pattern. Alternatively, there are the so-called guided waves which remain contained within the wave guide. For example, lamb waves are guided waves travelling along the plate. Rayleigh waves are guided waves which are constrained to the surface. Love waves are guided waves which travel in layered materials. Stoneley waves are guided waves, constrained to material interface. The ultrasonic NDE essentially relies on the elastic wave propagation and its reflection within the material. These wave field disturbances are caused due to local damage and any other defects that are present. On the other hand, any disturbance caused to this wave field is an indication of the presence of damage. Ultrasonic waves involve the following measurements:

(1) time of flight (TOF), wave transit or delay in wave transmission;
(2) path length;
(3) frequency;
(4) phase angle;
(5) amplitude;
(6) acoustic impedance;
(7) angle of wave deflection (or refraction).

There are different techniques of ultrasonic measurements.

- pulse echo method;
- pulse transmission or pitch-catch method;
- pulse resonance method.

Procedure

An ultrasonic probe (commonly used is a piezoelectric one) which is placed on the surface of the element inducing waves in the material.

These probes establish contact with the structure using special coupled gels. Now, based on the incidence of the transducer with respect to the surface of the member, it generates P, S waves or its combination. These waves detect anomalies around the sound path. In the pitch echo method, defects are detected in the form of echoes. In the pitch-catch method, flaws are detected by the wave dispersion and its attenuation due to damage.

Major drawbacks of using ultrasonic method

- The sound path traverses only on the small portion of the material volume. Hence, the transducers should be moved to cover a large column which is time consuming. Alternatively, one can use C-scans, but they are expensive.
- Ultrasonic waves cannot be induced normal to the surface of the structure. Hence, cracks that are normal to the surface cannot be detected readily by this process. But, guided waves can be used to detect such flaws. For example, lamb waves are commonly used guided waves. They are used to detect faults or damages in sheet metals, air frames, large containers and pipes. Lamb waves can detect cracks, intrusions, disbanding in metallic and composite structures. They are very useful for detecting damage in thin plates and shells. However, Rayleigh waves are more useful in detecting surface defects.

2.29.2.   *Embedded NDE*

Sensor network used for monitoring can be completely embedded (permanently fixed) into the structure. These sensors can be used for monitoring. There are two ways by which this can be done:

(i) *Passive SHM*

Passive SHM which uses passive sensors that are monitored over a period of time. The monitored data will be useful in updating the system characteristics, for example, load sensors, sensors to measure stress, environment conditions, acoustic emission from cracks, etc.

Passive SHM only listens to the structure. It does not interact with the structure.

(ii) *Active SHM*

It detects the presence of damage and also estimates its extent and severity. One of the active SHM methods is the piezoelectric wafer active sensor (PWAS). These sensors send signal which are essentially lamb waves and also receive lamb waves to identify the presence of damage in the structure (cracks, delamination, de-bonding, corrosion, etc.). In flat plates, ultrasonic guided waves travel as lamb waves and shear horizontal waves. Lamb waves are vertically polarised, while shear horizontal waves are horizontally polarised.

Consider a plate with stress-free upper and lower surface as shown in Figure 2.5. The local axes are marked in the figure. The thickness of the plate is '2d'. The free body diagram of the small area extracted from the plate is also shown in the figure with the reference axis and the corresponding stresses marked.

The equation of motion for an isotropic elastic motion is given by

$$\mu\nabla^2 u + (\lambda + \mu)\nabla\nabla u = \rho\frac{\partial^2 u}{\partial t^2},$$

where (λ, μ) are lane's constants, ρ is the mass density and u is the displacement vector. Assume displacement vector as

$$u = \nabla\phi + \nabla \times H,$$

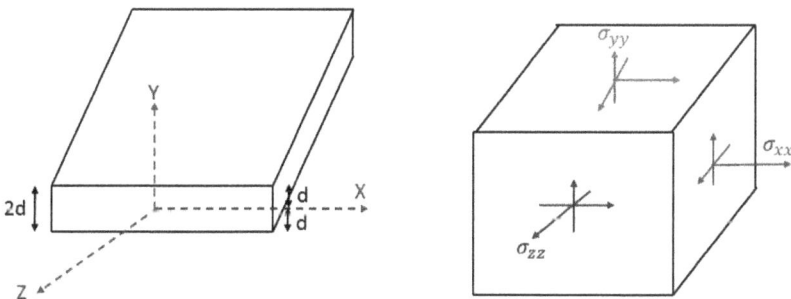

Figure 2.5. Plate with free boundary.

where ϕ, H are potential functions given by

$$\phi = f(y)e^{i(\xi x - \omega t)},$$

$$H = (h_x(y)i + h_y(y)j + h_z(y)k)e^{i(\xi x - \omega t)},$$

where ω is the circular frequency, ξ is the wave number and wave speed, C is the ratio of circular frequency and wave number. The governing equation is now written by

$$\nabla^2 \Phi = \frac{1}{C_p^2} \frac{\partial^2 \Phi}{\partial t^2},$$

$$\nabla^2 H = \frac{1}{C_s^2} \frac{\partial^2 H}{\partial t^2},$$

$$\nabla \cdot H = 0.$$

For the plane strain, which is z invariant, the above equation reduces to the following form:

$$f'' - \xi^2 f = -\frac{\omega^2 f}{C_p^2},$$

$$h_x'' - \xi^2 h_x = -\frac{\omega^2 h_x}{C_s^2},$$

$$h_y'' - \xi^2 h_y = -\frac{\omega^2 h_y}{C_s^2},$$

$$h_z'' - \xi^2 h_z = -\frac{\omega^2 h_z}{C_s^2},$$

where C_p which is a pressure wave speed longitudinal component and it is given by

$$C_p^2 = \frac{\lambda + 2\mu}{\rho},$$

$$C_s^2 = \frac{\mu}{\rho},$$

where C_s is the shear wave speed transverse component. The solution of this equation will lead to the following:

$$\phi = (A\cos\alpha_y + B\sin\alpha_y)\,e^{i(\xi x - \omega t)},$$
$$H_x = (C\cos\beta_y + D\sin\beta_y)\,e^{i(\xi x - \omega t)},$$
$$H_y = (E\cos\beta_y + F\sin\beta_y)\,e^{i(\xi x - \omega t)},$$
$$H_z = (G\cos\beta_y + H\sin\beta_y)\,e^{i(\xi x - \omega t)},$$

where $\alpha^2 = \frac{\omega^2}{C_p^2} - \xi^2$, $\beta^2 = \frac{\omega^2}{C_s^2} - \xi^2$, A–H are constants which can be determined from the stress-free boundary conditions at both upper and lower surfaces of the plate. The characteristic equations are given as follows:

$$-A(C_3\sin\alpha d) + H(C_4\sin\beta d) = 0,$$
$$A(C_1\cos\alpha d) + H(C_2\cos\beta d) = 0,$$
$$B(C_1\sin\alpha d) - G(C_2\sin\beta d) = 0,$$
$$B(C_3\cos\alpha d) + G(C_4\cos\beta d) = 0,$$
$$-E(C_5\sin\beta d) + D(\beta^2\sin\beta d) = 0,$$
$$-E(\beta\sin\beta d) + D(i\xi\sin\beta d) = 0,$$
$$C(\beta^2\cos\beta d) + F(C_5\cos\beta d) = 0,$$
$$C(i\xi\cos\beta d) + F(\beta\cos\beta d) = 0,$$

where

$$C_1 = (\lambda + 2\mu)(\alpha^2 + \lambda\xi^2),$$
$$C_2 = 2i\mu\xi\beta,$$
$$C_3 = 2i\xi\alpha,$$
$$C_4 = \xi^2 \quad \beta^2,$$
$$C_5 = i\xi\beta.$$

It can be seen that there are pairs which are formed. These are called coefficient pairs of this characteristic equation. The pairs are (A, H), (B, G), (E, D) and (C, F). (A, H) and (B, G) correspond to symmetric and non-symmetric lamb waves, respectively. (E, D) and (C, F) are symmetric and non-symmetric shear horizontal waves, respectively. For each characteristic equation, one can find the specific value of the wave number and wave speed, which gives the solution for this equations.

Embedded sensors

The guided waves can be excited by impinging the surface with ultrasonic beam in oblique angle. This can be induced by a large ultrasonic sensor or ultrasonic transducer fixed at the wedge (Table 2.7). This can generate a combination of pressure waves and shear waves into the structure. They can also be alternatively created by comb transducers. Comb spacing tunes the guided waves to its half wave length. In the recent past, researchers have also used piezoelectric wafer sensors (PWAS) to generate guided waves. The advantages of these sensors are as follows:

1. They are very light in weight (60 mg).
2. They are very cheap.
3. They are very simple and thin (0.2 mm thick).
4. They are unobstructive to the surface where they are embedded.

These sensors provide bidirectional energy transduction from the device to the structure and receive it back from the structure into the device. They operate on the piezoelectric principle that couples the electrical and mechanical variables in the material. Let the mechanical strain be S_{ij}, mechanical stress be T_{kl}, electric field be E_k and electric displacement be D_j, then the following equation holds good:

$$S_{ij} = S_{ijkl}^E T_{kl} + d_{kij} E_k,$$
$$D_j = d_{ijkl} T_{kl} + \varepsilon_{jk}^T E_k,$$

where S_{ijkl}^E is the mechanical compliance measured at zero condition, $E = 0$, ε_{jk}^T is the dielectric permittivity measured at mechanical stress $T = 0$ and d_{ijkl} is the piezoelectric coupling effect.

Table 2.7. Comparison of embedded and ultrasonic sensors.

Ultrasonic sensors	Embedded sensors
The conventional ultrasonic sensors are weekly coupled because they are connected to the structure through gels	Embedded sensors are connected to the structure permanently because they are embedded inside the structure
These sensors are resonant narrow banded type	These sensors are non-resonant broadband type. They can be tuned for a wide range of frequency of certain lamb waves
They sense lamb waves indirectly through acoustic waves by impinging them on the surface	These sensors excite lamb waves directly through in-plane coupling

Procedure of PWAS

The piezoelectric effect converts stress applied to the sensor into an electric charge. Similarly, the converse piezoelectric effect produces strain when voltage is supplied to the sensor. These PWAS can act both as exciters and detectors of elastic lamb waves travelling in the material. They can be used as both active and passive probes.

Applications of PWAS

- They can be useful in active sensing of far-field damage using the pulse echo procedure, pitch-catch method and phased array method.
- They can also be useful in active sensing of near-field damage with high frequency impedance method.
- They can be useful in passive sensing of crack initiating and location by the acoustic emission method.
- They can also be useful for passive sensing of damage through low velocity impact detection technique.

Based on the incidence of the transducer with respect to the surface, waves may be created. It may be P waves, S waves or their combination. These waves detect anomalies around the sound path they travel. In the pulse echo method, defects are detected in the form of echoes. Alternatively, in the pitch-catch method, flaws are detected by wave dispersion caused due to damage.

(iii) *Embedded NDE using pitch-catch method*

This method is suitable for embedded NDE. This method can be used to detect the structural changes that take place between the transducers. One transducer will be placed as a receiver and another transducer will be working as a transmitter. Pitch-catch method can detect changes that are created by guided wave amplitude, phase difference between them and their dispersion. This method has the following applications:

- corrosion detection in metallic structures;
- damage detection in composite materials;
- detection of de-bonding in adhesive joints;
- detection of delamination in layered composites.

In the embedded method of NDE, transducers are permanently inserted either between the layers of the composites or attached on the structure permanently. It is advantageous to detect cracks in metallic structures. Cracks in metallic structures generally form and propagate perpendicular to the surface. Generally, it can cover the whole thickness. In such cases, the crack is called through thickness crack. The consequence of such crack is that it can tear the metallic structure. In the conventional NDE, cracks in metallic structures are generally detected with ultrasonic or eddy current probes. One of the limitations of this method is that they can detect cracks or flaws only at particular points. If one needs to examine the crack presence of the whole surface, one should have to manually scan over the complete surface to detect cracks. This is a very tedious exercise and this has a possibility of over sighting a few crack locations. This can be corrected by the pitch-catch method.

In the pitch-catch method, guided waves are transmitted from one location and received in another location. Thus, the whole member or material is analysed for the following:

(i) guided wave shape and its amplitude;
(ii) phase lag created between the end signals;
(iii) change in amplitude.

These will help us to detect the presence of cracks without manually scanning the complete surface. This can be seen as one of the important advantages of the pitch-catch method of NDE when it is used for crack detection in metallic structures. This method can detect the presence of cracks and also their extension without scanning the whole surface manually. There are researches to ascertain that the probability of crack detection by the pitch-catch method is much higher than other methods. This is given by the following relationship:

$$P \text{ (crack detection)} = \text{Sum of cracks recorded}/((M - N) + 1),$$

where M is the number of crack events recorded by NDE method and N is the number of serial events. It is very important to note that the pitch-catch method is also effective in detecting fatigue crack propagation.

2.30. Crack Detection in Composites

Composites generally resist the loads by the layered structure. In case of formation of through thickness cracks, their propagation in composites are resisted by the presence of reinforcement fibres. Hence, cracks grow parallel to the surface especially at the interface between the layers. They are generally initiated by the following: (i) fabrication imperfection; and (or) (ii) inability to resist fatigue loads. On the other hand, in the conventional NDE, ultrasonic probes are used to sense additional echoes to capture these surface parallel cracks. P waves will be reflected by delamination of layers. This will be an indication of crack development parallel to the surface which essentially causes delamination in composites. Pulse echo method can also be used for crack detection in composites. In such case, an appropriate guided wave (lamb wave) must be chosen to detect the crack. It is seen that lamb waves show better reflection from a through thickness crack. They should be less dispersive. The advantages could be as follows:

- Better reflection ensures a strong signal for crack detection.
- Less dispersion ensures compactness and convenience to interpret.

In various places, wide applications of the pitch-catch method are seen in pipe lines, closed conduits (tubes) and cables.

2.30.1. *Embedded phase arrays*

NDE also uses another technique called embedded phase arrays. In this technique, real-time phased array systems are used. They have transducers to inspect very thick specimens, for example, reinforced concrete slabs of deck of a bridge with P waves.

2.30.2. *NDE: Time reversal method*

The signal sent by a transmitter arrives at the receiver, while it gets modified in the medium though which it travels. If the received signal is reversed and sent back from the receiver to the transmitter, then the effect of the medium through which the signal travels is also reversed. This technique is called the time reversal method. This is very useful when dispersive lamb waves are employed for damage detection. One of the exclusive applications of this method is usefulness in ultrasonic imaging of the difficult medium. By comparing the discrepancies between the original input signal and the reconstructed signal, damage can be detected.

Chapter 3

Sensor Technologies

This chapter deals with various sensor technologies that are commonly deployed in the health monitoring of civil engineering structures. A few sensors that are vital for specific measurements required in offshore platforms are also discussed. The sensor layout and details of the structural health monitoring scheme are presented with component-level details.

3.1. Sensor Technologies

The most frequently measured parameters in structural health monitoring (SHM) decide the types of sensors to be deployed. These parameters can be grouped into three types: (i) mechanical type that measures strain, displacement, deformation, crack opening, stress and loads; (ii) physical type that measures temperature, relative humidity (RH) and pore pressure; and (iii) chemical type that measures chloride and sulphate penetration, pH changes, carbonation penetration, rebar oxidation, timber decay, etc. Monitoring using sensor networking can be done at different levels. Therefore, sensors should be installed to appropriately measure the desired parameters at various levels of monitoring. Naturally, all sensors do not apply to all levels of monitoring. The levels of monitoring are as follows.

Very early stage of monitoring: In this case, embedded type sensors with low stiffness are used. Generally, such monitoring practices exist in concrete structures and are primarily used to study the shrinkage effects at the early stage of construction. It is also useful to measure strain that occurs due to extreme weather changes. In this case, the period of measurement can vary every hour for the first 24–36 hours; subsequently, interval of measurements may vary from a day to a maximum of a fortnight.

Continuous monitoring: In this type of monitoring, interval of monitoring varies from 24 to 48 hours; period of measurements can be 1 hour. Variations in the natural period of the structure caused due to temperature effects and load effects are continuously monitored.

During construction: This is very important in the case of structures expected to have foundation settlement effects and those structures are built near the fault lines of seismic signals. Based on the schedule of construction, the period of monitoring can be fixed. At least one measurement per sensor at each construction stage is necessary.

Testing stage: Certain structures are subject to test loads before they are actually put to functional use, such as railway bridges, highway bridges, reservoir structures, coastal jetties and dockyards. In such monitoring, period of measurement is one recording per sensor per stage of testing, where load change is significant.

Period before reconstruction: In this case, monitoring is done several times a day at irregular intervals. In addition, there can be a periodic monitoring, which is continuous over a period of 24–48 hours. This type of monitoring is carried out to determine the effect of temperature and load variations on the existing structure. Based on these observations, reconstruction is demanded.

During reconstruction: In this stage, period of monitoring can be four times a day per sensor for 24–48 hours. There can be many such sessions of recording during every stage of reconstruction.

Long-term monitoring: This is carried out during the service life of the structure. In this case, period of measurement can be at least one to four measurements per day per sensor. Further, it can be one measurement per sensor per week to one per sensor per month, which can be subsequently extended once in a year.

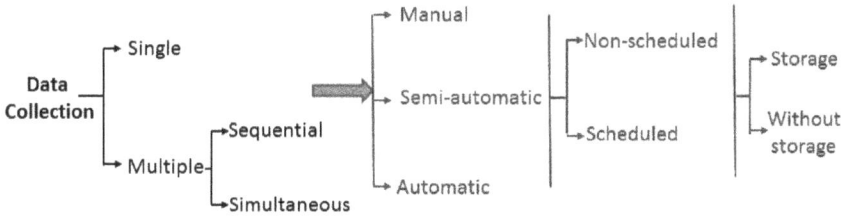

Figure 3.1. Data collection and management.

Special events: Measurements are also observed during the event (if possible) and after the event to understand the post-event damages and new damages, if any, that are initiated. Special events such as heavy rains, which can result in foundation settlement and extensive corrosion; strong winds, which cause foundation uplift, extensive bending stress and formation of flexural cracks; and during earthquakes. The type of sensors chosen for such monitoring depends upon the level of monitoring.

Data management: It is the management of measured data, which is very crucial. These data can be collected by various ways: manual; semi-automatic; and fully automatic. They can also be collected on-site or by a few remote application methods. It can also be collected periodically or continuously, either in static or dynamic modes. Further, it is important to see that these options can also be combined. For example, in the case of a coastal jetty, long-term monitoring can be combined with maximum (or minimum) performance observation; this can be either automatic or remote communication. If the data are continuously monitored, then it is preferred to handle the data without any human intervention. The data collection layout is shown in Figure 3.1.

3.2. Fibre Optic Sensors

The application of fibre optic sensors (FOS) is contrary to that of electric sensors in many ways. FOS use electromagnetic interference to read or measure data, whereas electric sensors use electric pulse. Due to the low-light attenuation of optical glass fibres, which are essentially used in FOS, these sensors can be used in several kilometres long application, whereas electric sensors have serious limitations in this front. The classification of FOS depends on various parameters:

(1) based on light characteristics (intensity, wavelength, phase or polarisation, etc.), which are modulated by the parameters to be measured;
(2) based on whether light in the sensing segment is modified inside or outside the fibre (intrinsic or extrinsic);
(3) based on whether they are local, quasi-distributed (or fibre Bragg grating — FBG) or distributed sensors (or Brillouin scattering distributed FOS);
(4) based on how are they installed; generally, they are surface-mounted, but sometimes, they can be embedded as well.

3.2.1. *Measurement of moisture ingression*

Moisture ingression is one of the major problems in buildings. The most important task is to measure this data. It is difficult to identify the source of moisture ingression and its path of propagation because it is a surficial phenomenon. FOS can be used to identify the location of moisture ingression. These sensors consist of a swelling-type, polymeric FOS, which are used to measure the distributed moisture formulation. This sensor works in combination with the optical time-domain reflectometer to determine the spatial location of moisture ingression. This measures (or identifies) the point of moisture ingression by attenuation principle. A typical sensor is shown in Figure 3.2. It consists of a protective layer of felt wick, which is connected by an optical fibre, which is embedded to optical

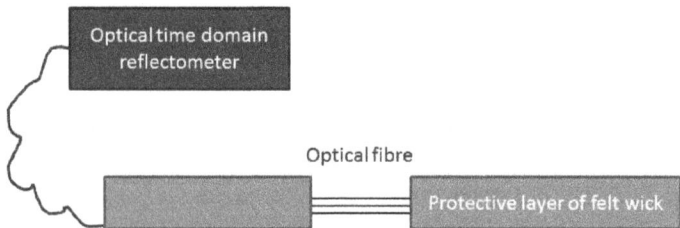

Figure 3.2. Typical FOS.

reflectometer. The device consists of an optical fibre, a polyvinyl alcohol hydrogen rod, which is embedded inside a protective felt. The device can sense the bending of fibres at microstress-level variations. Hydrogel has a characteristic of swelling in the presence of water; swelling without dissolution causes the optical fibre to undergo a microbending. The microbending of the fibre interferes with attenuation of light, which is transmitted through the fibre. Through this process, it measures the location of moisture ingression.

3.2.2. *Single-point RH sensor*

FOS are also commonly used as a single-point RH sensor, which measures the RH. This sensor is made of polyamide-coated FBG. Due to the wavelength, the encoded RH readings are measured by the sensor. FBG sensors are coated with polyamide to protect them from any external damage. Several such sensors can be used in parallel to measure the RH. The polyamide coating in FBG sensors acts as a hygroscopic coating that swells in the presence of water vapour due to the absorption of water molecules. This causes strain in the FBG sensor, which actually depends on the applied RH linearly. Now, by tracing the reflected Bragg wavelength, RH value of the location where the sensor is placed can be measured. This is very useful in tropical locations, where RH can cause corrosion to material. It is generally useful to plan preventive maintenance in the case of monumental buildings.

3.2.3. *Crack sensor*

This is used to locate the flexural cracks in beams and slabs of buildings. Optical fibres are integrated into a textile-net structure, which is designed to transfer elongation due to cracks developed on the structure to the optical fibres. Since the failure stress of optical glass fibres is relatively low, integrated optical fibre will break even under the formation of small cracks; optical time-domain recorder is used to locate the cracks. Integration of the optical fibre into the textile structure can be done in two ways: (i) by stitching the fibre to the structure, or (ii) by knitting. The principal objective of fabrication is to minimise losses due to bends and obtain the best bonding.

Alternatively, optical fibres are also used to monitor the shape of the crack tip in concrete members. The network of optical fibres breaks when cracks propagate in the members, thus making these fibres interesting. They are very helpful to locate the cracks. Alternatively, optical fibres can be laid in a zigzag manner at the bottom of the concrete beam to detect the flexural cracks in the beam. When the cracks open in the member, optical fibres, intersecting the cracks at an angle other than 90° had to bend. This sudden bending of the fibre causes optical power loss, indicating location of the cracks. This method is highly suitable to detect cracks of size smaller than 0.1 mm. Optical fibres should be laid inside the concrete member such that these fibres should be free to slide inside the concrete.

3.2.4. *Crack detection in composites*

Alternatively, there are sensors used to detect cracks in composites. This is also called as the self-monitoring technique. These FOS contain an electric conductive phase such as carbon fibre and the conductive power in cement or polymer mix. These sensors can monitor their own strain, damage and temperature variation effects by the embedded or reinforced carbon fibres, which act as self-monitoring sensors. Different composites, which act as self-monitoring sensors are listed in Table 3.1. They are all examined

Table 3.1. Different types of composites.

Composites	Electric conductive	Matrix material
Carbon fibre-reinforced concrete (CFRC)	Short carbon fibre ($L < 10$ mm, $V < 0.5\%$)	Cement, mortar, concrete including admixtures
Carbon fibre-reinforced polymer (CFRP)	Short carbon fibres ($L < 10$ mm), continuous carbon fibres	Resin, curing agent
Carbon fibre glass fibre-reinforced polymer (CFGFRP)	Continuous carbon fibre ($V < 0.5\%$)	Resin, curing agent
Carbon powder disbursed in glass fibre-reinforced plastic	Graphite carbon powder ($V = 0.15\%$, average particle diameter = 5 microns)	Resin, curing agent

only in the lab scale. A real-time application is yet to happen in the sense of its large surface areas.

3.3. Magnetostrictive Sensors

Ferromagnetic materials when placed in the magnetic field are mechanically deformed. This property is called *magnetostrictive* effect. In the reverse, magnetic induction of material changes when the material is mechanically deformed. This is called inverse magnetostrictive effect. These sensors are useful to detect voids in concrete-filled steel pipes. These sensors could generate guided waves of different modes propagating along the length of the pipe. These waves are then sensitised to the defects in the pipe and they detect the damage. The received wave amplitude decreases with the increase in voids and inclusions; this indication is useful to detect damage in concrete-filled pipes. One of the major disadvantages of this sensor is that the ultrasonic energy emitted is very low in strength; this can be improved by combining the sensor with other piezoelectric sensors.

3.4. Smart Sensing for SHM

Sensors with embedded microprocessor and wireless communication links are called smart sensors. These are very useful in wireless sensor network (WSN). This is the most updated system in SHM scheme. The advantages of smart sensors are as follows:

(1) It has an ability to continuously monitor the integrity of the structure in real time.
(2) It can provide improved safety to the public, particularly in the case of ageing structures like bridges.
(3) It has an ability to detect damage at an early stage, which can then reduce the cost of repair and minimise the shutdown time of the structure.
(4) It is helpful in predicting or observing initiation of damage or any other undesirable behaviour of the structure, e.g., settlement of supports, fatigue formation, etc. Therefore, they can be helpful in generating advance warning of removal of the structure or declaring it non-functional due to safety regulation. So,

essentially, it can prevent serious disasters, which are actually caused by structural damages.

Smart sensors are more or less wireless. In conventional wired sensors, there are many numbers of wires. Sometimes there can be fibre optic cables or physical transmission medium, which may be a serious issue in the case of long-span bridges or tall buildings. On the other hand, wireless sensors have low cost and densely distributed network. Recent advancements in wireless sensors like wireless communication, microelectromechanical systems (MEMS), advanced information technology to enhance the SHM quality are positive features of wireless SHM. Sensors are also available nowadays with self-calibration and self-diagnosis capabilities. Essentially, sensors have three components:

(1) sensing element, consisting of a resistor, capacitor, piezoelectric module and photodiode;
(2) signal condition processor consisting of amplifiers, linearisation, compensation and filtering;
(3) sensor interface, which includes wires, plugs and sockets to communicate with the other electronic components.

There is an essential difference between the smart sensor and the conventional sensor: Smart sensors have a microprocessor on board, which makes them intelligent. Microprocessors can perform the following functions:

(1) digital processing;
(2) analog-to-digital convertor (ADC) or frequency code convertor;
(3) calculation of interface functions, which can facilitate self-diagnosis, self-identification and self-adaptability or decision making;
(4) it enables to control storage of data or dumping of data;
(5) more importantly, it can decide when to remain in operation and when to go in sleep mode. This can save power to a larger extent.

3.5. MEMS Sensors

MEMS sensors are manufactured using a very large-scale integrated technology, termed as VLSI. This leads to manufacturing sensors in large quantity and reduces the cost of the sensors. These sensors perform the integration of mechanical and electrical functions. Sensing operation requires a physical (or a chemical) phenomenon to be converted into an electrical signal. This is useful for display, processing, transmission or even recording. Alternatively, these sensors can also be used as actuators in terms of control analogy. Actuators reverse the flow. They convert electric signal into a physical action (or a chemical change) in the system. Size of MEMS sensors are very small; they are about 10^{-6} m. it is interesting to note that the mass production of these sensors may bring down the cost and enable their use in SHM widely. All sensors are essentially wireless if they are to be categorised as smart sensors. They also possess capacity with data transmission based on radio frequency (RF) communication. Smart sensors have four main features:

(1) They have on-board central processing unit (CPU).
(2) They are small in size and compact to use.
(3) They are wireless and hence there is no congestion due to wiring.
(4) Low cost can be achieved if mass production is enabled.

Since SHM using smart sensors are mostly automatic systems, the primary requirement of such systems should be enabling preventive maintenance when there is a likelihood of the response exceeding the threshold value. SHM using smart sensors involve a five-level classification:

(1) It will first assess the response and determine whether the structure is damaged or not.
(2) If damaged, it further shall try to locate the damage which is called damage localisation.
(3) Based on the data observed or monitored, it shall quantify the damage (extent of damage).
(4) It shall also predict the future progress of damage and remaining service life of the structure.

(5) Finally, it should recommend appropriate remedial measure or repair measures to restore both strength and functionality of the structure.

SHM using smart sensors offer a complete solution for safety and healthy functionality of the structure.

3.6.   Sensing Requirements in Special Structures

Structures, which are of strategic importance like naval base, monumental structures, offshore platforms, nuclear power plants, reservoirs, etc., need to be monitored for various reasons. There are about 1470 oil rigs located offshore and about 7000 platforms, which perform oil and gas drilling around the world in various geographical locations. These platforms have topside components, which include living quarters, helipad, drilling equipment, electrome-chanical equipment, cranes, etc. A typical size of a platform is about 90×90 m in plan, which is quite a massive, huge-spanned structure. They have heavy mass concentration spread over a large area. This is a congestion, which causes accident in offshore platforms. In addition, offshore platforms have huge capital investment apart from habituating 100–150 people working (or residing) on board. These people are specially trained technicians and engineers, whose manpower is highly valuable. Their downtime, resulting from any repair, could stop production and cause loss of revenue. Further, it attracts a variety of loads, such as wave load, wind load, current load, ice load, seismic load, impact load, dead load, live load, machinery load, vibration load and accidental load. Moreover, it stocks highly inflammable material (oil and gas), which is the primary source of accident.

From the statistics, major accidents in offshore platforms occur due to explosion, loss of structural integrity and fire; consequences of such accidents are very severe. It results in severe damage to the structure and poses threat to both the environment and human life. Offshore platforms handle hazardous chemicals like petroleum products, oil and gas, which have the potential to cause major accidents. For example, Alexander Kieland semi-submersible platform, which capsized in March 1980, resulted in the death of 120 people. Ocean ranger oil drilling rig accident occurred on 15 February 1982 in North Atlantic Sea off coast of Newfoundland, Canada resulted in serious economic loss; 84 crew members died. Mumbai High North disaster that occurred on 27 July 2005 in India caused the death of 22 persons. The Bohai-2 oil rig disaster in November 1979 in the Gulf of Bohai, China, resulted in a major setback in the economic situation. The major hazard in all the above accidents is the flammable condensate and its leakage. This is essentially due to poor or delayed maintenance, improper planning and avoidance of preventive maintenance. The essential solution must have been a continuous monitoring of certain parameters, such as temperature, pressure, etc. Wired sensors cannot be employed because of congested layout and their complicated network. Smart sensors are essentially required under such situations. One of the common factors found to be interesting in the above accidents is the lack of communication between the maintenance staff and the operation staff. Delay in the maintenance schedule, inadequate maintenance and not following safety procedures could add to this complexity.

3.7. Sensor Requirements and Data Acquisition

Offshore structures operate under high-risk factor. This is essentially due to the kind of process which they undertake in terms of oil exploration and production. They need to be monitored because they are novel in their type and the topside mechanical systems are usually custom-designed. The most important issue is their failure or downtime for repair, which can cause a significant economic loss. They are all strategically important. Therefore, the primary requirement of such a structure is preventive maintenance approach, which is essentially dependant on continuous monitoring of the structure under time-varying loads. It is difficult to carry out inspection traditionally through non-destructive tests or even through visual inspection (VI). The basic reason being that the structure is huge, partially submerged in sea, and certain areas or members of the structure cannot be inspected. An automated continuous SHM scheme is necessary because then the damage analysis can be carried out to ensure operational and functional safety. In areas where VI is not possible (piles, foundation members), one should examine these members through simulated numerical model. A few scaled model of the same platform can be examined experimentally. A correlation need to be established between the observations made at the lab scale and that of the prototype. Certain approximations are adopted during the experimental investigations of offshore structures to make the procedure more convenient. They are as follows:

(1) Varying mass is not linked with the marine growth, equipment and fluid storage (which otherwise is temporary).
(2) Variable submergence, leading to change in buoyancy and mass of the members, is not included. This will influence the energy dissipation of the system significantly.

As suggested by Brinker *et al.* (1995), there are certain factors that govern the design of monitoring system for offshore platforms. They are as follows:

(1) Sensors should be able to withstand environmental uncertainties.
(2) Proposed SHM scheme should have financial advantages over the manual inspection method which is more or less traditional.
(3) The vibration spectrum should remain stable over a period of time.

Table 3.2. Vibration-based monitoring sensors.

Physical parameter	Principle of the sensor	Technology
Acceleration, velocity, displacement	Inductive sensors, capacitive sensors, piezoelectric sensors	Conventional, MEMS technique
Magnetic field, magnetic resistivity	—	Magnetoresistance meter
Optical properties	Photoelectric sensors, optical fibre sensors	Fibre Bragg grating, Fabry–Perot interferometer, intensity-based sensor
Acoustic type	—	Ultrasonic probes

(4) Normal sea state and wind excitation should be used to extract the natural frequency of the system.

(5) Above water measurements should be used to identify the mode shapes.

Table 3.2 summarises the characteristics of sensors used in vibration-based monitoring.

3.8. Sensor Performance

The quality of data in health monitoring depends on the performance of the sensors. Some of the common factors that govern the selection of sensor are as follows:

- data format;
- precision and accuracy;
- linearity of the data;
- dynamic range of the variables;
- cross-talk;
- durability;
- maintainability;
- redundancy;
- cost of the structure.

Thus, in offshore structures, choice of sensor, investment on the type of sensor and networking are not influenced by the economic perspective as long as the necessity of SHM of an offshore platform is well established. SHM involves various activities like detection and tracking. During detection, sensor is prepared to read the data and correlate the data with the sensitivity of damage. There are two approaches by which this can be done. The most common approach is the deployment of sensor or array of sensors in a network with commercially available components in the market. The major disadvantage is that excitation of the structure will be limited to the range of frequency of the array of sensors. Physical quantities are measured without any definition of damage. The assumption here is that the measured data will be sensitive to the damage. It also has an assumption that the damaged and undamaged structures are subjected to a similar kind of excitation, which is a very serious limitation. The same strategy is employed in real time, which will now measure the data and analyse the data for damage-sensitive features.

Alternatively, quantify the damage before developing the sensing system by numerical simulation. The model is prepared through numerical simulation and results available from the simulation prior to the occurrence of damage are noted. They will provide the type of damage, the possible extent of damage and the location of damage. The vital parameter to design the sensing system is that the extraction of damaged information features from the data. One can use statistical pattern recognition (SPR), which is now the governing factor to design the acquisition system. Based on the outcome of SPR, additional requirements, which arise from the environmental and operational conditions, are updated. The operational conditions are helpful to predict the initial detection of damage, which can be done by numerical simulation. This improves the quality of the damage detection process in SHM. The acquisition depends on the type of data to be handled. This should be defined to design the sensor network system. There are mainly two types of data: kinematic quantities and environmental quantities.

3.9. Acquisition System and Networking for SHM

The type of data that need to be monitored should be defined to design the sensor network. There are two types of data: kinematic quantities and environmental quantities. Kinematic quantities include displacement, velocity, acceleration, strain measurements, etc.; one can use the traditional type of sensors to measure these quantities, e.g., accelerometer (both uniaxial and triaxial), LVDT, strain gauges, displacement transducers, force transducers, load sensors. However, to measure environmental quantities, such as temperature variation, pressure, moisture content, RH, one needs a special dedicated kind of sensors. But, these parameters not only affect the damage level of the system but will also have an impact on the operation of the sensors. WSN is one of the advanced options in terms of smart sensor. It is very useful for automatic and continuous monitoring. The advantages include reduction in price, simple installation and affordable network in comparison to wired network. Further, wireless systems have low system cost (networking cost). They have very low installation time. In addition, one can avoid the congestive layout of wires. So, there are no complexities that may arise from their laying and in-service maintenance, which are common problems with wired sensors. Further, wired sensors depend on a central server to communicate, whereas wireless sensors do not need a central server. They convert measured data into digital form and transmit them directly. Therefore, WSN makes online monitoring more simple, low cost with low-cost processor to handle the data. There are many advantages of SHM applied to strategic structures:

- It ensures serviceability of the structure through long-term monitoring.
- It increases safety and knowledge about the performance of the structure.
- It validates the design of the structure and its performance.
- It can monitor and control the construction process during operation as well.
- It can assess the load capacity and therefore the risk of the structure.
- It can assess any requirement of emergency response efforts.

WSN essentially originates from wired network with the centralised data acquisition (DAQ) unit. Sensors are used to measure physical parameters as analogue values. Then these sensors are connected to a centralised data server through wires. Then the DAQ converts analog-to-digital signals and then processes the data. Wired sensors are expected to give high-quality measurements and input to the SHM scheme. There was no transmission delay and no data loss. But, all these were true when done on the lab scale. On a real-time monitoring, there were many problems. Some of the important issues include dependency or reliability of sensors or wires for a long run, power exploitation, network congestion, and failure to work on demand. So, they cannot be implemented on large structures, such as bridges, dams, offshore platforms, etc. If the cable is damaged, data will be lost and this will lead to loss of efficiency of the whole system. The electromechanical systems which induce machine vibrations, cross-interference of power signals with that of the sensors created additional complexities. Wireless sensors are essentially smart sensors. They have self-adaptability and self-scalability to monitor on their own. Thus, wired sensors do not have the capability to process data. So, the centralised server is responsible for collection, aggregation, storage and processing of data. The whole concept of data management is centralised, which is negative in wired sensors. This disadvantage can be avoided by using wireless sensors.

3.10.   Wireless Sensor Networks

Wireless sensor networks (WSNs) have many overruling advantages. WSN is considered as an alternative for wired network. Wireless sensors eliminate the need for physical power and data cables. This reduces the complexity in layout. This also reduces the cost of network and the installation time of network. The most important advantage is that the amount of data that can be measured by the monitoring systems which employ WSN is phenomenally high. So, WSN (smart sensors) has the capacity for information extraction, data processing and data compression. They eliminate the need for a centralised server which was a weaker link in the wired system. Wireless sensor nodes also have a microprocessor in each sensor node, which can process the data and filter the data based on any previous input, which reduces the volume of data to be transferred. The most important advantage with wireless sensors is that they measure the data or value as soon as damage is identified. It means that there is no time delay in real-time monitoring in acquiring the data, processing the data, and in deciding whether the data are necessary or not.

In 2009, sensors with integrated management system were developed. This used the ubiquitous computing technique. The design enabled both TCP and IP protocol. The fabricated wireless unit measures and transmits the data through Bluetooth technology. Ambient vibration measurements were made using these sensors and were found to be very efficient. This system had an acquisition unit, which collects the data from various sensor nodes of different types. It enabled measurements from all sensor nodes simultaneously. The unit was tested experimentally and found to be very efficient. This was used in Golden Gate Bridge to measure the vibration without interfering its operation. This used MEMS sensors with high accuracy and high reliability. It can also be embedded in the structure with the capacity to measure the damage at an early stage to design and alert the monitoring system. This was first proposed by Chung *et al.* (2004). The sensing unit was designed with RISC microcontrollers and MEMS-based accelerometers. The sensing unit showed fast computation capability for data processing.

The major issue is related to SPR. In the SHM scheme, a module was developed, which can do operational evaluation, data acquisition, feature extraction and then a statistical model. Hence, new forms of transmission boards are developed by coupling the integrated hardware approach (see, for example, the transmission board developed by Motorola). One of the main advantages is the ready availability of wireless accessing point in the sensing unit itself. The first integrated DAQ system has telemetry and processing system for the SHM scheme. The new development came into play with the use of imote2, which was designed by Intel. This provides a powerful computational tool and powerful communication platform which were in highest demand in SHM application in the recent past. The board has signal processing unit and high-resolution DAQ. These sensors are very useful in large application, commissioned in test ship MV OOCL, Europe. SHM scheme measured the hull stresses with the wave monitoring system. The data were validated with detailed experimental and numerical analysis and found to be very efficient, and the comparison is reported by Yu *et al.* (2006).

3.11. SHM Layout Design for Offshore Structures

One of the major challenges of SHM design in offshore platform is the non-stationary response of the offshore platforms. There is a continuous change in mass called added mass and a continuous

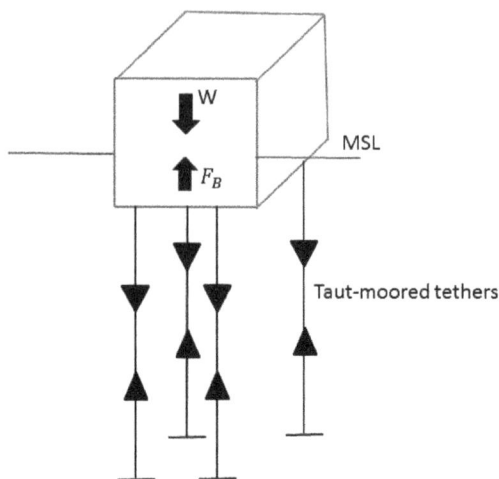

Figure 3.3. Compliant platform.

change in stiffness characteristics. The study is carried out on a compliant structure. Compliancy means flexibility. The conceptual model of compliant platform is shown in Figure 3.3.

The basis for the design of offshore structures is that the buoyancy force should be greater than the weight. Thus, to hold down the platform, taut-moored tethers are used. The equation for static equilibrium of this design could be as follows:

$$T_o + W = F_B.$$

The difference between buoyancy and weight is balanced by tension. When the tethers slacken, this changes the stiffness of the platform. These platforms undergo major structural modifications due to the encountered environmental loads, which occur from wave, wind and current. Sometimes, it challenges the safety of the platform. Under this condition, the requirement could be that the response of the platform should be within the permissible limits, so that all operations like production, storage, and transportation can be carried out safely. Thus, the major factors include production activity, safety and serviceability, which should be monitored. VI is found to be a boon in many damage identification philosophies. They were done successfully through VI. Unfortunately, VI is not possible in offshore structures because the members under water are inaccessible, the platforms are located in hostile environment and the structural characteristics change continuously. Thus, automated monitoring system is required, which could be carried out on a continuous basis and should notify only when the characteristics change significantly. Above all, the monitoring system should be simple, self-diagnosed and autotuned to the platform. An automated real-time interpretation is very important. There should not be any packet loss, data overflow, malfunctioning of sensors and adaptability. The different features such as threshold crossing, modal parameter identification and structural degradation should be of importance. The method available in the literature for offshore structures is global vibration-based damage detection. If the damage is due to some catastrophic effect, the damage cannot be detected because of marine growth. There is a poor visibility of the platform. There is a need to identify damage only by detecting change in significant characteristics, which needs demand of assessment of offshore platforms. Changes in properties need to be compared with undamaged platform, then a detailed numerical modelling of the platform and its response behaviour under environmental loads should be available. It is required to compare the damage case with the undamaged case to find the changes. A thorough numerical modelling is essential. Generally, frequency domain approach is convenient to do this. One of the major challenges in SHM design of offshore platforms is the location of sensors. Sensors placed close to the damage site are influenced more than those placed away from the site. The most important issue is the distribution of sensors, so that it detects the damage without fail.

Following are some issues with respect to vibration-based damage detection:

- noise measurement and signal-to-noise ratio;
- discrepancy between scaled-down model and prototype;
- nonlinearity in the structural response;
- dense distribution of sensors;
- influence of environmental factors in real time, which cannot be considered in the lab scale.

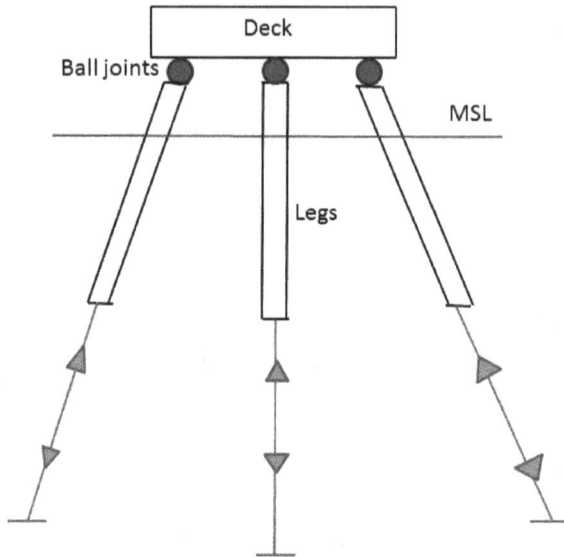

Figure 3.4. Conceptual model of BLSRP.

The application of this concept to a new generation platform called buoyant leg storage and regasification platform (BLSRP) is explained in the following. The conceptual model of BLSRP is shown in Figure 3.4. The deck is connected to six buoyant legs, whose position are restrained by tethers. The buoyant legs are connected to the deck by hinged joints, which do not transfer rotation from the leg to the deck and vice versa. But they transfer translations completely. This platform has six degrees of freedom in both deck and buoyant legs. The ball joint transfers surge, sway and heave degrees of freedom and restrains roll, pitch and yaw degrees of freedom. Thus, this platform is stiff in the vertical plane and compliant in the horizontal plane. They are essentially used for storage and regasification. This is one of the recently developed novel concepts in the design of offshore platforms.

BLSRP has many complexities which arise because of the loading condition, geometry and design. Wireless sensors are used to monitor the response of BLSRP. To perform this, failure is initiated in the platform and then the damage is detected. WSN is developed based on vibration-based monitoring. The performance of the platform is examined by measuring strain, displacement, force, and acceleration. The displacement is one of the major factors which control the design of any system. One is interested to identify the level of damage and location of damage in relation to displacement and drift. So, the platform undergoes dynamic behaviour, damage resulting from the members may also have a different duration and bandwidth of acceleration. Now, the primary idea is that the acceleration signal measured from a damaged model is compared with the undamaged model to know the significance of damage. Since the study is done on the lab scale, extreme wave loading under which the damage can occur is modelled using endurance wave analysis approach. Peak frequencies of PSD functions are considered, which in turn reduces the time history record. So, the design of SHM should be capable of recording or monitoring the relative motion between the wave and the platform.

3.12. Components of SHM System

The hardware components of SHM system consists of four functional modules:

(1) sensor unit;
(2) DAQ unit;
(3) computational core;
(4) wireless communication channel.

The most important commonness between these modules is that, in all modules, commercially available hardware components will be used. The technical details of these components will be examined for its versatility, availability and cost. The primary objective of the SHM design is to arrive at a layout of sensors costing lesser and functionally efficient. The components of SHM systems are shown in Figure 3.5.

3.12.1. *Sensing unit*

These are the fundamental blocks of SHM design. Sensor-based SHM system measures two quantities:

- kinematic quantities (acceleration, inclination, strain and displacement);

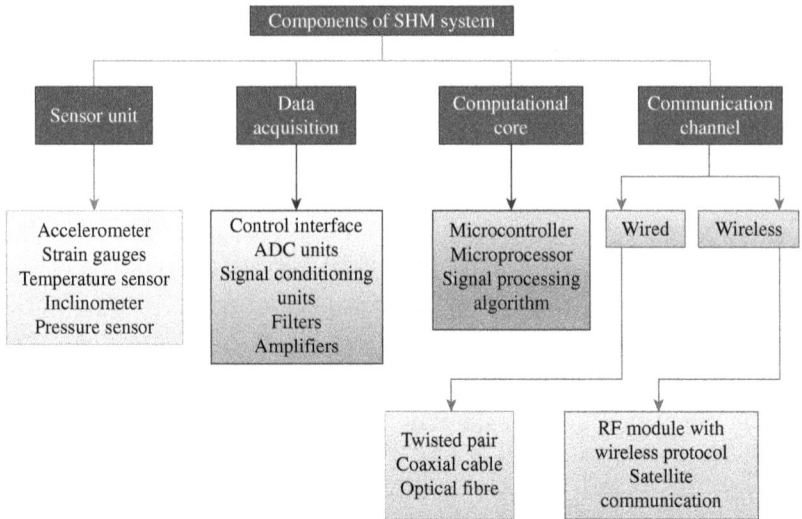

Figure 3.5. Components of the SHM system.

- environmental quantities (temperature, humidity, wind direction and speed).

Wireless sensors are supposed to monitor both types of quantities. It also enforces the level of monitoring. Offshore structures are massive, and thus the level of monitoring required should be ensured. Sensor technology has advanced forward as sensors, wireless sensing unit and smart sensing unit. One can use the combination of these based upon certain basic criteria:

(1) capability to capture both local- and system-level responses;
(2) capability to capture or acquire data in both consistent and retrievable manner for long-term data processing and analysis;
(3) adaptability to operational conditions and the hostile environment.

We must choose sensors that are commercially available in the market to reduce cost if the production is massive. Thus, both the sensing requirements and the instrumentation capability should be ensured.

3.12.2. *Data acquisition unit*

It also physically measures the response. It has a hardware to measure or receive the response. It also has a PC with a required software to process the received signal in a useful manner, may be in a graphical form or tabular form. The basic parameters to be considered for choosing the data acquisition (DAQ) unit are as follows:

- sampling rate;
- resolution of the data;
- number of channels;
- type of transducers.

In the present design, commercial model SPIDER-8 is used with the Catman Express software which processes the data. Wireless modules are chosen such that they match with the DAQ capacity and that of the wired ones. It is very important that the sensing interface should be designed properly so that it remains compatible with all types of sensors and for an appropriate sampling rate. The traditional SHM systems have the following:

- analogue sensors;
- ADC at DAQ end.

Most of the sensors are MEMS-based as they occupy less space. These sensor modules also have an inbuilt ADC. The additional important factors include the number of sensors, type of sensors, spacing between the sensors and DAQ which controls the data degradation possibility, noise filters and low-pass filters.

3.12.3. *Computational core*

It will be a microcontroller unit which has inbuilt algorithms to interrogate the measured data. It is not a blind analyser, but it is an intelligent source of producing only meaningful output in the desired format, which is the requirement of the day. Therefore, computational core is a combination of data storage, processing unit, which needs very high memory to store the measured data and also process it in the desired format. The minimum requirement of the system could be 256 kB RAM memory in the lab scale. So, data are stored and processed only on a temporary basis. The data should be transmitted as early as possible and it is not going to act as a permanent data storage system. The processor consists of expandable memory of SD cards, operating systems, USB flash drive and interface for external devices. It is also programmed to perform customised operations, such as destined data collection, storage, processing and transmission, as demanded by the SHM system. Here, retrieval is also an important stage.

3.12.4. *Communication channels*

The two operating channels are wireless channel and wired channel.

(i) *Wireless channel*

The design parameters which govern the design of wireless channels are as follows:

(1) data rate;
(2) open space range;

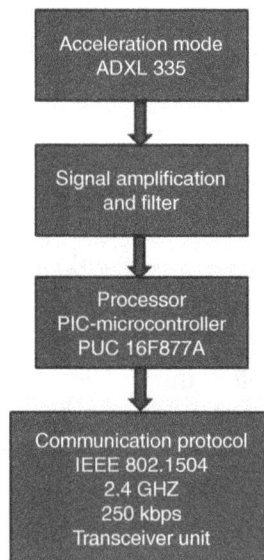

Figure 3.6. Wired communication systems.

(3) encoding reliability;

(4) radio band.

They are designed based on the required functionality. The main parameters in choosing the channels could be reliability and range. There are adverse possibilities of data loss due to interference, reflections, and path loss, which need to be addressed in the WSN design architecture.

(ii) *Wired channel*

It has got an acceleration mode which essentially is ADXL335. It has a signal amplification and filter unit which takes the signal to the processor, as shown in Figure 3.6. The PIC microcontroller PIC16F877A is used.

3.13. Artificial Intelligence in SHM

Artificial intelligence (AI) plays an important role in SHM. SHM is a large and complex file with many scientific and practical aspects. The SHM system is a heterogeneity of various engineering technologies. The major aim of SHM is to accurately identify the current state of health and behaviour of the structure. This can be generally achieved by automatically analysing the measured data (monitored data), which are received from various monitoring devices. There will be anomalies during measurement, which should be detected and as far as possible eliminated to obtain the state of health of the structure in real time. It is also necessary to evaluate the structural deterioration and damages precisely. The greatest advantage of doing this is that the maintenance cost of the structure can be greatly reduced. We can aim towards preventive maintenance and avoid downtime cost and reduce the duration of shutdown time for repair. The service life of the structure can also be enhanced. Here, AI plays an important role.

AI has a strong genesis in computer science. It provides a variety of methods for monitoring problems which would be otherwise difficult to solve. It is computationally capable of solving complex problems. AI incorporates human-like intelligence, covering a similar thought process, consciousness, and self-awareness, and it replicates

Table 3.3. Comparison of AI.

Conventional AI	Computational Intelligence
It has symbolic school of thought	It has sub-symbolic school of thought which also includes neural networks
It represents human knowledge explicitly in a declared form	This incorporates human thinking at sub-symbolic level
They implement procedural knowledge and expertise which are transferred into them through training and data simulation generated by symbols and symbolic structures	By modelling the mental phenomenon, elementary units are interconnected in the network and knowledge is implicitly represented
Examples: Expert systems, case-based reasoning, Bayesian network	Examples: Neural networks, fuzzy logic systems, evolutionary computation

the biological models which lead to the definition of intelligence. AI also enhances special computational capabilities solving mechanisms and algorithm which actually simulate the intelligence of human behaviour without including any direct relationship to human abilities. This is totally a mechanically set automated procedure which thinks, acts, decides and controls the situation similar to human intelligence. The conventional AI and computational intelligence are compared, as shown in Table 3.3. The basis for comparison is with respect to human intelligence.

SHM has certain levels of decision making. It also has complexities of wrong or incorrect data. It is called latency problems. These problems in SHM can be handled by AI effectively. The process of SHM will be separated from the overall software system to individual microcontroller units, intelligent sensors which are located on the structure. AI transfers or decentralises the complexities from software system to sensing systems. The main purpose of sensing units is to automatically control the acquisition of data. Sensors are not simply recording the data. They are now meant to control the acquisition of data. This shift is a result of AI through smart sensors. This can be expressed graphically, as shown in Figure 3.7.

The acquired or measured data may have inconsistency. They may not be compatible with the preset threshold values. It is necessary to assess the errors in the data using microcontroller-based

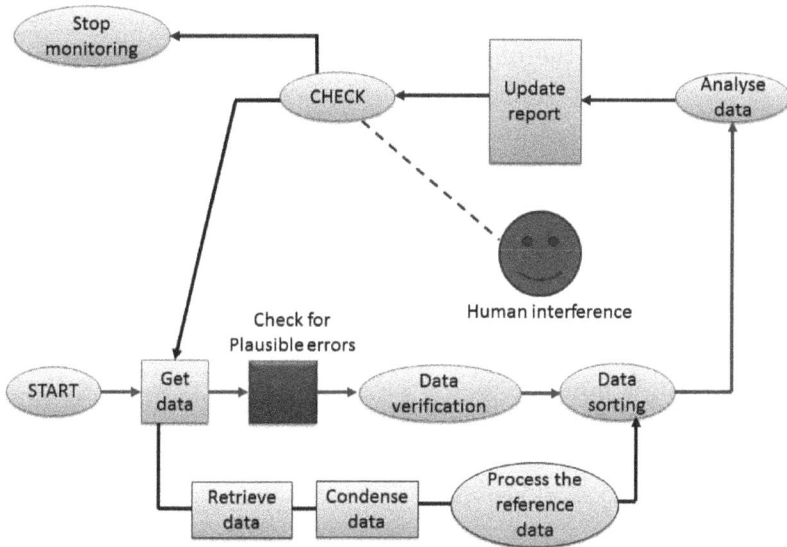

Figure 3.7. AI in SHM.

sensing units. These units will compare the acquired data with the previous set of data. If no significant change is seen for a long period of observation, then the sensors will interpret this as wrong data. For example, if equal values are reported from a particular sensor location repeatedly for 'n' times, this is marked as inconsistency. Errors of this nature can also be identified by performing regression analysis done by the sensor. This detects plausible errors in the acquired data. This is an important step which makes AI different from the conventional SHM. One can use a simple microcontroller with limited computational power to do this job. If the response measurement (y) of a structure at a particular sensor location is recorded as (y) and predicted values lead to (y_p). By comparing the predicted values with the acquired values, plausible errors can be detected. If $|y - y_p| > \delta y_p$, then the permissible range δy_p will be a function of the measured and predicted values. It depends on the kind of project of SHM. Therefore, the predicted values can be given as follows:

$$y_p = \beta_0 + t\beta_1 + \varepsilon,$$

where t is the time index used to compare y and y_p, β is the regression coefficient and ε is the unpredicted or unexplained variation in the predicted value. Depending upon the sensor type, the error function can be automatically condensed into the output. So, the unexplained errors are eliminated.

The data analysis can be done either by short-term or long-term analysis. Short-term data analysis has two steps:

(1) prognosis;
(2) evaluation.

The prognosis value is computed using the simple multiple regression model:

$$(y_p)_s = \beta_0 + x_1\beta_1 + x_2\beta_2 + \cdots + x_k\beta_k + \varepsilon,$$

where the parameters x_i are the corresponding variables independently measured from different sensor locations. Now, based on the prognosis value, the measured variable can be evaluated by the fuzzy logic approach. The long-term data analysis is generally dealt with using data mining and machine learning technique. Further, both the analyses can lead to a successful pattern recognition which makes SHM more a closed-form problem. So, data trends which are different can be handled by Mann–Kendall test to check for the possible pattern.

3.14. Artificial Neural Network in the SHM Process

There are four axioms which are very useful and directly applicable in the case of AI in SHM.

Axiom III: Identifying the existence and location of damage can be done in an unsupervised learning mode. Identifying the type of damage present in a system and severity of the damage can be done only by a supervised learning mode.

Axiom IVa: This states that sensors cannot measure damage. It is only the feature extraction, done through signal processing and statistical analysis, that classifies the damage from the sensor data.

Axion IVb: Without intelligent feature extraction, changing operational conditions and environmental data makes the measured data of damage more sensitive.

Axiom V: The length and timescales associated with damage initiation and evolution decide the properties and characteristics of the health monitoring system.

The intelligence in SHM can be useful in composite structures. Thus, the main emphasis is on using a robust-type signal processing protocol. For example, glass fibre-reinforced plastic (GFRP) laminates are generally used as structural materials because they have a high strength-to-weight ratio and good corrosion resistance. They are also useful in military applications because they minimise electromagnetic radar signature on underwater vehicles. Under static and dynamic loads, they fail mainly due to cracking or delamination. Delamination is more severe because it causes stiffness reduction and leads to catastrophic failure of the structure. Therefore, it is vital to detect the delamination in GFRP. A more vital part is that the limited delamination may be invisible, but still they can cause severe damage to the mechanical properties and load-carrying capacity of the structure. The techniques available to check delamination are as follows:

(1) X-ray;
(2) ultrasonic C-scan;
(3) laser shearography.

The difficulty with these methods is that it takes considerable time to inspect the GFRP structure by using these techniques. The desired option could be online detection of damage. Artificial neural networks (ANNs) combined with preprocessing tools, such as damage relativity analysis technique (DRAT) can be used for damage diagnosis. This can predict the location, size, presence and extent of the damage precisely.

ANN is a large parallel distributed process comprising simple processing units. These units are called neurons which have multiple interconnection points. ANN are capable of mapping the relationship between measurable features of structural damage to the physical parameters. The classification and identification of structural damage can be successfully done using ANN. It uses a set of known damage features and their corresponding physical parameters. It also employs multi-layer feed-forward backpropagation network to perform data segmentation, data compression and pattern recognition if present by identifying the repetition of data.

ANN applications are largely seen in bridge structures. For example, consider a railway bridge whose health monitoring is required to be done. The steps involved includes the data collection from the dynamic response of the bridge through simulation under the passage of train. When this is done, it is assumed that the bridge is in undamaged state and it is therefore considered to be healthy. This can be done in two different damage scenarios. In the first stage, one can use ANNs which are essentially trained with an unsupervised learning approach. The input comprises the accelerations of the deck under a healthy state. Now, based on the acceleration value at the previous instant of time, the neural network predicts the future acceleration. In the second stage of damage prediction scenario, the prediction errors are statistically characterised by a Gaussian process which supports the choice of damage detection decision from a known threshold value. By comparing the damage indices with the threshold value, one can differentiate the health conditions of the bridge. For each damage case scenario as seen above, operating characteristic values in the form of curves are obtained. Then, using Bayes theorem, one can also estimate the total cost of the proposed methodology.

The damage detection can be done model-based or model-free. In the first case of ANN, one needs to have a computational model of the sample structure. In this case, it is a railway bridge. The damage detection through this first scenario will have a direct physical interpretation. But, it will be difficult to develop a highly accurate numerical model of a complex structure. The second case actually deals with the model-free approach. By means of AI, the model is classified or identified, and tried to relate to the physical characteristics of the bridge. The model-free approach consists of training an algorithm on some sample acquired data, usually it is done in an unsupervised manner. Without a detailed numerical model, using AI damage is identified and correlated to physical characteristics based on the algorithm of sample data. In ANN, one of the vital issues is the placement of sensors. One needs to look for an optimal sensor placement, which is the deciding feature of successful damage detection. People generally use genetic algorithms to choose the sensor type and their location. Dual structure coding and mutation particle swarm optimisation (DSCMPSO) algorithm is recommended. In this case, the convergence speed of the algorithm is highly improved and the optimal location is also highly improved.

The next challenge of using ANN in SHM could be the separation of changes in structural characteristics that are caused by vibration-induced damage and changes in operational and environmental conditions. The solution for this problem could be the use of Kalmar filter which can be applied to the network for damage detection essentially caused by temperature changes. In this method, the Kalmar filter is used to estimate the weights of the neural network and the confidence intervals of the natural frequencies which are used for damage detection. ANN can be successfully used in SHM, which is dependent on the machine learning algorithm (MLA). Generally, MLA is implemented to detect structural abnormalities caused by the monitoring of data. They use outlier principle to detect these abnormalities. It is based on the training data, which is exclusive to a problem being solved.

ANN actually uses data-driven approaches for decision making. They are successful in civil engineering structures because of the following reasons:

- Large quantity of sensor data are available for civil engineering structures.
- Physical characteristics of the structure are complex to model.
- Computational efforts of ANN need to be reduced.

In such situations, the option used by researchers is decentralised ANN. It follows an embedded machine learning approach to perform the autonomous detection of sensor failures. Therefore, data analysis in SHM is related to transforming the useful compact sensor data into useful information, probably in the knowledge format so that it can be related to the physical characteristics of failure such as life cycle prediction and life cycle management. Two approaches are data-driven and physics-based, which can be used in ANN as applied to SHM system. The physics-based approach establishes the principle model and maps the physical characteristics; and then compares the output of the physical model and finally decides the damage. The data-driven approach is similar to the unsupervised and untrained data system. It is also computationally intensive. This approach depends on the comparison of observed data with the previously collected sensor data to decide the damage scenario. These are useful when the large sensor data are available and physical characteristics of the structure are modelled.

Chapter 4

Applications of Structural Health Monitoring*

This chapter deals with the use of various types of sensors and their limitations. Factors influencing the design of layout of sensors along with the advantages of wireless sensor networking are discussed. Preliminary designs of both wired and wireless sensor networks for health monitoring of tension leg platform and buoyant leg storage and regasification platform on lab scale are presented. Structural assessment of a Bust duct system and supporting pipe racks for electric transmission lines are also investigated for their seismic safety by imposing a postulated failure. Details of structural assessment, as part of structural health monitoring, are presented.

4.1. Introduction

Infrastructure development is inevitable for the growth of world's economy. This, in turn, strongly depends on existing structural and mechanical systems, such as buildings, aircrafts, bridges, machineries, offshore oil rig platforms, ships, power generation units and many more. All these systems show high degree of uncertainty in their behaviour under the encountered environmental loads. In order to ensure safety during operation and survivability during extreme conditions, it is imperative to detect the state of health of these structures under progressive damage caused, if any.

---

*This chapter has been co-authored by Dr. Thailammai Chithambaram, Visiting Research Fellow, University of New South Wales, Sydney, Australia.

Offshore structures are used for the exploration of oil and gas under the sea. There are about 1470 offshore rigs and 7000 fixed and compliant offshore platforms in the world. Offshore structures are huge and massive, which run over large span with heavy mass concentration spread over the entire area.

Offshore structures operate under high risk factor due to the kind of process involved in exploration and production. Apart from being novel and expensive of its kind, their failure may lead to serious environmental pollution and economic loss as well. To avoid such serious consequences, it is better to follow preventive maintenance approach, which shall depend on continuous monitoring of the structure under time-varying loads. In case of offshore structures, it is difficult to carry out traditional inspection through non-destructive testing or visual inspection as the structure is huge and partially submerged in the water while manual inspection is not possible in all locations. An automated monitoring, such as the use of sensors in structural health monitoring (SHM), will be an effective tool to assess the status of the platform based on the damage analysis to ensure safe operability.

Offshore platforms need a huge capital investment for their installation before they start their operations to earn revenue. Considering a limited population of high-tech personnel living on-board and custom-designed plants and equipment installed on board, their downtime could cause a huge financial loss. It is therefore imperative to monitor their structural conditions on a continuous (or at least on intermittent) basis to ensure both satisfactory functionality and public safety. Offshore structures are subjected to various types of loads, such as permanent or dead load, operating or live load, environmental loads (wind, waves, earthquakes), ice loads, temperature loads and marine growth, accidental loads and construction loads. All of them possess a high degree of uncertainty in terms of magnitude, direction and duration of application. While wind load acts on one portion of the platform, which is located much above the draft level, wave loads constitute a major component of the lateral loads. Marine growth accumulated on the submerged members contributes to adding up to the mass of the structure apart from obstructing access to carry out visual inspection and non-destructive evaluation. It is therefore interesting and equally important to assess the behaviour of the platform under extreme wave loads to ensure its ability for withstanding storm loads during critical sea-state condition. This is done in view of public safety as a primary concern; a secondary concern is the economic survivability.

In oil and gas platforms, major accidents occur from explosion, loss of structural integrity, fire, etc. They result in severe damage of the structure in addition to posing environment threat and loss of human lives. Offshore platforms handle hazardous substances, such as petroleum, chemical, oil and gas, which have the potential to cause major accidents. Risk is implicitly present in offshore oil and gas exploration, which is not eliminated but limited to as low as reasonably practical (ALARP) level. Hence, it is imperative to monitor the health of offshore platforms to avoid serious and catastrophic accidents. Major hazard is the flammable condensate and its leakage, which would arise due to delayed maintenance schedule and poor maintenance planning.

4.2. Health Monitoring of Offshore Structures

Some of the major factors that influence accidents in offshore structures are due to poor maintenance, lack of communication between maintenance and operation staff, delay in scheduled maintenance, inadequate maintenance and safety procedures (Chandrasekaran, 2016a–d; Okoh and Haugen, 2013). Damage scenarios are due to various reasons, which include ship impact on the structure, fatigue and damage due to corrosion. Researchers have simulated the damage either by inducing a crack or by removing the member. Experiments are performed on the scaled platform tested in lab to detect the self-induced damages (Begg *et al.*, 1976). In real-time implementation of health monitoring, an attempt to do measurements below the water line will be expensive and could be highly rational as divers are employed for measurements. Alternatively, measurements below water line can be done using underwater sensors with wiring, for which the wiring has to be done at the time of construction of the platform. In such cases, corrosive saltwater environment is an important factor to be considered for long-term monitoring. In case of the offshore structure, as most of the areas are not accessible for measurement, damage assessment scenarios are generally examined through simulated numerical models while a few on the scaled models are examined through experimental investigations. It is seen as a good practice to analyse change in response of scaled platforms under different loads during experimental investigations and extrapolate the data for failure analyses or correlate these data with those measured on the real platform.

While deploying SHM in offshore structures, certain assumptions and approximations are important for a convenient procedure as follows: (i) varying mass is not linked with the marine growth, equipment and fluid storage; and (ii) variable submergence, leading to change in buoyancy and mass of structural members is not included and it will alter the energy dissipation of the system (Brincker *et al.*, 1995). Based on the studies reported in the recent past, factors that govern the design of monitoring system for offshore platforms are as follows (Loland and Dodds, 1976):

- sensors should withstand environmental uncertainties;
- proposed SHM scheme should have financial advantages over the traditional (manual) inspection method;
- vibration spectrum should remain stable over a period of time;
- normal sea state and wind excitation shall be used to extract the resonance frequency;
- above water measurements should be used to identify mode shapes.

4.3. Vibration-Based Monitoring for Offshore Structures

Vibration-based damage detection techniques are commonly used in damage diagnosis as they are one of the most efficient methods. Among various sensor technologies, the microelectromechanical system (MEMS) has its own advantages and is applied in various fields, such as biomedical, automotive, construction and consumer sectors. Vibration-based damage detection methods are further classified into traditional and modern approaches. Traditional method is based on the principle that change in mass and stiffness will be reflected in the measurements of natural frequency and mode shapes of the structure. When the measured data of natural frequency or mode shape are different from that of the normal, it indicates the initiation of damage. Modern type is the online measurement of structural response to detect damage with the help of signal processing techniques, artificial intelligence and neural networks. Dynamic response of the structure under different loads is measured online while SHM indicates the change of structural parameters thereby detecting damage in the structure.

4.4. Sensor Performance and Data Acquisition

Quality of data in health monitoring system depends on the performance of the sensors. Some common factors that govern are data format, precision and accuracy, linearity of data, dynamic range, cross talk, durability, maintainability, redundancy, calibration and its cost. SHM involves detection and tracking. While the first step is to make the sensor reading correlated with the sensitivity of damage, tracking involves establishing relationship between the damage features and damage levels. In general, there are two approaches in which the SHM system is developed. The most common strategy for developing the sensor network for SHM is to deploy an array of sensor network with the commercially available components. Excitation of the structure is limited to the range of frequency of these arrays of sensors. Physical quantities are measured without any definition of the damage that has to be detected, with an assumption that these measured data will be sensitive to the damage (Farrar *et al.*, 1994). This is based on the assumption that damaged and undamaged structures are subject to similar kind of excitation. Such strategy is deployed in the real time, which will measure the data and analyse the data for damage sensitive features. Alternate strategy is to quantify the damage through some process before developing the sensing system. Based on prior availability of numerical simulation results, damage location and type of sensors are chosen. Extraction of damage features and statistical pattern recognition will be a part of the data analysis, which will be vital in the development of data acquisition (DAQ) system (Flynn, 2010). Additional requirements are updated based on the changing environmental and operational conditions. The latter one with the initial prediction about the damage by numerical simulations improves the probability of damage detection.

Sensors are devices that detect and measure a typical input from the physical environment, such as temperature, displacement, heat, pressure, etc. They are inevitable components that are already part of our everyday life. For example, a modern car uses more than a hundred sensors. Apart from measuring a particular quantity, other properties of sensors are also as important as sensitivity. Sensors should also be sensitive to just the desired measurement. Further, sensor properties should remain unchanged to draw conclusions from the measured signal to the stimulus to be measured. A few of the important properties of sensors other than sensitivity, selectivity and (long-term) stability are sensor response, resolution and measurement uncertainty. In SHM, sensors will be operated for long periods of time and under challenging environmental conditions. In addition to this complex work environment, there is often the demand on a high level of accuracy. Even though the commercial market is full of a variety of sensors, it often takes many years of research and development to identify the appropriate type of sensor and design its layout for SHM.

Types of data that need to be acquired should be defined to design the sensor network system. Two major types of data are (i) kinematic quantities; and (ii) environmental quantities. Kinematic quantities include displacement, velocity, acceleration and strain. Traditional types of sensors are used to measure these dynamic responses. Accelerometers, displacement transducers and force transducers like load cells are some of the sensors used in SHM. Environmental quantities include temperature, pressure and moisture content. These parameters not only affect the damage level of the system but also have impact on the operation of the sensors.

4.5. Sensor Networking

SHM originated from the wired sensors with the centralised DAQ units. It has the sensors which can measure the physical parameter as the analog value; these sensors are connected to the centralised data server through wires. The DAQ system converts the analog signals to digital signals and then processes the data. This type of SHM with wires will give accurate measurements without any data loss and there is no transmission delay. But this holds good for a laboratory environment. In real-time implementation of large structures, such as bridges, dams, offshore platforms, this will lead to many complexities as it has to run extensive length of cables. If there is any damage in the cable, it is difficult to find the discontinuities throughout the cable length. Due to high-level complexity of electromechanical systems and their interferences with the structural geometry, monitoring through wired sensors results in inefficiency as their failure within their connections becomes untraceable. The implementation time and the cost of wired system are also expensive. Wired SHM systems have no means to locally process their data; rather, the centralised data server is responsible for the aggregation, storage and processing of all measurement data.

SHM is a typical field in which the application of the wireless sensor network (WSN) is useful to both measure damages and online monitoring. Due to the reducing price and advancements in recent technologies in sensor networking, it is now easy, simple and affordable to have WSN in lieu of the traditional wired SHM. Wireless SHM reduces the system cost and the installation time. In addition, it eradicates the installation of lengthy cables and thereby reduced the complexities involved in their laying and in-service maintenance. While wired system depends on the centralised server, wireless nodes do not rely upon a central server. They convert the measured data into digitised form and transmit them directly. Wireless SHM makes online monitoring simpler with low-cost computing processor. Recent innovations in the wireless SHM lead to the migration of computational power from the centralised DAQ system to the sensor nodes.

After the research and experimentation process, the researchers have come up with WSN as the cost effective method for SHM — an alternative to the existing wired sensors. The SHM using WSN eliminates the need for physical power and data lines, thus reducing the complexity and the cost associated with installation, and the total system costs much less than the wired SHM system. The most important factor to be considered is the amount of data measured by these monitoring systems. In case of wired SHM, the DAQ is done by the centralised server, and the sensor unit will just capture the data and send it to the DAQ system without any processing and with no reduction in data based on the information it has. Wireless sensor nodes will have microprocessors in the sensor nodes which can process the data and filter based on the information in it, thus reducing the volume of data. The use of wireless communications to transfer sensor measurements to a centralised DAQ server is successfully illustrated (Straser and Kiremidjian, 1998). The wireless sensor node is shown in Figure 4.1. A dual microcontroller design was proposed with a low power 8-bit microcontroller, in which one is responsible for simple DAQ tasks and the other is used only for implementation of demanding numerical algorithms. (Lynch *et al.*, 2004).

In 2009, researchers developed an integrated management system using ubiquitous computing techniques. This design enabled the transmission control protocol/Internet protocol (TCP/IP) in which the wireless sensor unit measures the data and transmits it through the Bluetooth technology. Ambient vibration technology is adapted to identify the dynamic properties of the system (Heo and Jeon, 2009). Researchers developed an integrated network system which addresses some of the technical issues. This system enables real-time DAQ from various sensor nodes, which are capable of measuring the data simultaneously. It also has a low cost signal conditioning unit. The wireless MEMS-type sensor was developed to investigate the accuracy and reliability for real-time seismic monitoring. With embedded microcomputing units, it has the capability to detect the damage at its early stage by analysing the frequency response and gives warning to the user (Chung *et al.*, 2004). The wireless sensing

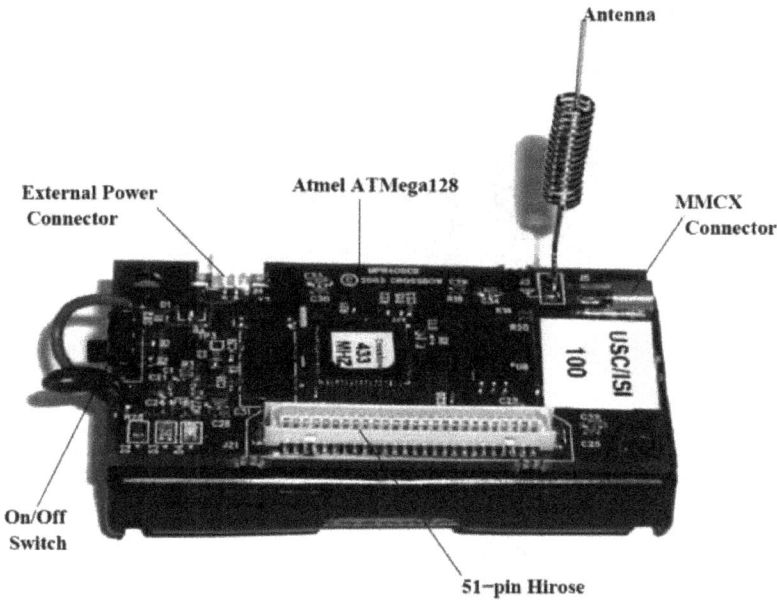

Figure 4.1. Wireless sensor node.

Figure 4.2. Wireless sensing unit.

unit is designed with RISC microcontrollers and with MEMS-based accelerometers. The sensing unit is capable of fast computation for data processing. The sensor node is tested with various experiments and some noise has been incorporated with the signal (Lynch *et al.*, 2001). The developed sensor unit is shown in Figure 4.2.

The issue faced in the SHM is approached with the statistical pattern recognition paradigm. In this, the entire process of SHM is

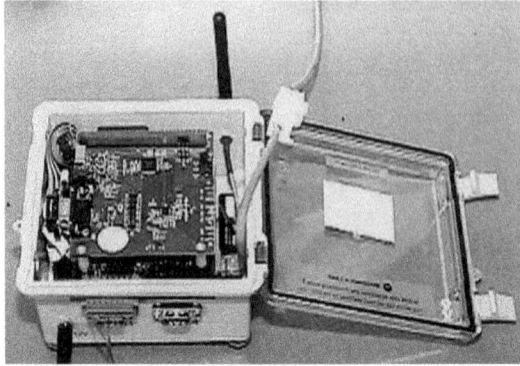

Figure 4.3. Wireless SHM system.

grouped as operational evaluation, data acquisition, feature extraction and statistical model development. This follows an integrated hardware software approach to get the solutions. This study discusses the coupling of data interrogation software with a wireless sensing unit. In particular, it will address the need to develop an integrated hardware software approach to get the SHM solutions. The transmission board developed by Motorola, neuRFon^{TM}, is the wireless access point for the wireless sensing unit. The image of the developed wireless SHM system with the transmission board is shown in Figure 4.3 (Farrar *et al.*, 2006). This is the first integrated data acquisition, telemetry and processing system for SHM. This shows better processing capabilities than the other embedded SHM systems.

The iMote2 shown in Figure 4.4 is designed and developed by Intel, provides powerful computation and communication for the demanding SHM application. It explains the development of the board with signal processing unit and accelerometer board for high resolution DAQ. The preliminary study shows that the combined analysis of acceleration and strain will result in more efficient damage detection (Rice and Spencer, 2008). The SHM system has been designed and developed based on the requirements of the large container carrier. It is installed in the test ship MV OOCL Europe from Busan to Jeddah. The SHM system consists of hull stress and wave monitoring system. The data

Figure 4.4. SHM board stacked on iMote2.

Figure 4.5. Narada wireless sensor node.

obtained assist in validating the analytical response of the ship motion, and derive the relation between the ship motion and sea loads (Yu *et al.*, 2006). The Narada wireless hull monitoring system shown in Figure 4.5 is developed and installed in the FSF-I Sea Fighter. The strain and acceleration response of the ship is measured using the Narada wireless sensor unit. It has 20 sensor nodes with acceleration channels, the data from these acceleration units assist to arrive at the deflection shapes of the ship. It emphasises that the future study should involve the embedment of data interrogation algorithms to allow the monitoring system to process its own data (Lynch *et al.*, 2009). Successful applications of WSN in the recent past for SHM encourage the attempt on an offshore platform (Daniele and Roberto, 2011; Mollineaux *et al.*, 2014; Peng *et al.*, 2009; Perez *et al.*, 2011; Wang *et al.*, 2006).

4.6. SHM in Offshore Structures: Challenges

One of the important challenges of SHM application to offshore platforms is their non-stationary response with a continuous change in mass and stiffness characteristics. Influenced by the encountered environmental and other loads that occur during processing and drilling operations, offshore platforms undergo major structural modifications that may even become structurally unsafe (Chandrasekaran and Koshti, 2013). The response of the structure must be acceptable for demanding requirements, such as production activity, safety and serviceability of the offshore structure (Komachi *et al.*, 2011). Manual inspection is not feasible on offshore platforms on a continuous basis; the automated monitoring system helps in identifying the noticeable changes in the response under severe environmental loads and also improves the operational safety. The automated real-time interpretation provides effective damage identification which includes various features, such as threshold crossing, modal parameter identification, etc. It identifies certain forms of structural degradation (Brownjohn *et al.*, 2011).

Vibration-based damage detection at the global level is essential for massive structures like offshore structures to localise it effectively. This is due to the fact that these are huge structures with their failure as a catastrophic event. Further, damage is poorly visible due to marine growth; also, environmental loading due to wave, severe storms and harsh environment affects the integrity of the platform. The changing platform condition and usage emphasise the need for assessment. The sensitive frequency domain for damage detection is identified for Jacket platforms in this study by the minimum rank perturbation theory (Kianian *et al.*, 2013). One of the important challenges for deploying SHM in offshore structures is the choice of location of sensors; sensors placed close to the damage site are influenced more than those located away from the site. Hence, dense distribution of sensors throughout the structure will detect the damage efficiently and make localisation easier. The common issues associated with the vibration-based damage detection listed in the NSEL report (Rice and Spencer, 2009) are as follows:

• noise measurements and signal-to-noise ratio;
• discrepancy between the scaled down model and the prototype;
• nonlinear structural response;
• dense distribution of sensors;
• influence of environmental factors in real time is not considered in experimental studies.

Details of preliminary experimental investigations carried out on a scaled model of offshore tension leg platform (TLP) are presented. Appropriate selection of wireless sensors and their networking geometry for DAQ are highlighted (Chandrasekaran and Thailammai, 2016; Chandrasekeran *et al.*, 2015). The monitoring experiments using wireless sensors are performed in an offshore structure model of Bohai Sea JZ20-2MUQ, which proved that the measured data reflect the vibrations of the platform. This study validates the feasibility of applying WSN to the structural monitoring of offshore structures (Yu and Ou, 2008). Performance of the offshore structure is analysed by measuring strain, displacement, force, acceleration, etc. Displacement is considered as one of the key parameters in the design of offshore structures. Based on the performance objective, the level of damage is identified and related to the displacement and drift. While variation in displacement shall be due to the change in water depth, environmental loads and damage caused by these loads, this alone will not control all performance objectives (Chandrasekaran, 2015; Chandrasekaran and Jain, 2016; Chandrasekaran *et al.*, 2017). Envisaged damage will also result from the duration and bandwidth of the ground motion, acceleration level, etc. In this study, acceleration and displacement are considered for health monitoring.

Response of the structure is captured by measuring the acceleration and inclination values in all dominant degrees of freedom, while the damage level is measured by computing the dissimilarities between the signals obtained.

The scaled model of fixed offshore structure is assessed under extreme wave loading using the modified endurance wave analysis approach. In this approach, it considers the effect of peak frequencies of power spectral density (PSD) and reduces the time history of the record (Dastan *et al.*, 2014). Numerous studies have been done on the offshore platform to analyse its response due to the environmental loads (Elshafeya *et al.*, 2009; Geng *et al.*, 2010; Idichandy, 1986; Li *et al.*, 2008; Park *et al.*, 2011). In addition to all these, the conversion of the acquired data into useful information for damage analysis of the structure remains a significant issue (Li *et al.*, 2016).

Offshore compliant structures are designed to undergo large displacement. The relative motion between the platform and waves counteracts lateral loads encountered by compliant structures; for large reduction in the net loads, compliant structures are designed to undergo large displacements in the horizontal plane but are restrained in the vertical plane. Therefore, unlike fixed platforms, stresses in members will be limited but displacements are intentionally kept larger. These compliant structures are suitable for both deep and ultra-deep waters. Some of the classic examples of compliant offshore structures are TLP, SPAR, Triceratops, buoyant leg storage and regasification platform (BLSRP), etc. In this study, acceleration responses of the platform in various degrees of freedom are measured in lab scale. Recorded measurements are analysed to identify damage-sensitive features, if any.

4.7. Methodology of SHM Design

Methodology of the SHM design shall be done in two stages. First stage is the design of the SHM system. In this stage, the hardware components are selected based on the requirement of the offshore structure model. The WSN technology is chosen for the experimental setup and the architecture is proposed. While threshold values are determined based on the response history of the platform under normal operating conditions, webpage and user interface are developed for post-processing and to display their current features. In the second stage, which is an experimental phase, the developed SHM system is installed in the scaled experimental model. Experiments are carried out for various postulated failure cases. Responses acquired using various sensors are processed in both time and frequency domains to highlight the damage sensitive features. Alert messages are triggered using the indigenously-designed alert monitoring system (AMS).

The current research in SHM is coupled with the miniaturisation of the electronic systems, hardware cost reduction, energy harvesting techniques and data processing methods. The success of these SHM technologies is mainly based on many factors: (i) extraction of damage sensitive features; (ii) large database management system; (iii) built-in data processing techniques in the sensor nodes; (iv) minimal environmental and operational variabilities; and (v) capability of the system to detect damages. Therefore, the classification of SHM technologies is based on the evolution of the sensors and monitoring processed over a period of time. SHM technologies in early stages are dependent on (i) wired sensor systems; and (ii) MEMS sensors with the DAQ system. However, the present SHM technologies are circumscribed to WSNs and extensive use of fibre optic sensor (FOS). Future SHM technologies are expected to focus on nanosensors, such as carbon nanotube (CNT) sensors and SHM, using artificial neural networks.

4.7.1. *SHM technologies in early stages*

(a) *MEMS with external DAQ unit*

The monitoring is then carried out with the miniature-sized sensors termed as MEMS sensors. Initially, these sensors were only meant for collecting the data; further transmission of data to the central DAQ system was not the part of their design. They measure physical quantity and send the raw data without any further processing.

MEMS sensors exhibit compactness in terms of their shape and size, reducing their component-level cost but fail to possess an inbuilt DAQ system to process the acquired data. They also do not possess scalability, which enables to connect them to nearby sensors and control/guide their performance. They act as standalone units connected to the central server.

(b) *Fibre optic sensor*

In general, the optical fibre has three layers. The innermost layer is the fibre core, which is made of silica glass and is surrounded by claddings. The core layer guides the transmitted light, whose refractive index is higher than that of the cladding region; this helps to confine the light within the core. The outermost layer is the Jacket, which protects the core and cladding from any external damages. Optical sensors are devices in which the input signal introduces modifications in some of the light characteristics. The device gives an output signal, which is processed and conditioned as a valid reproduction of object variables. The light can be modulated by varying its characteristics, such as amplitude, phase, frequency or polarisation state. The optical sensor becomes an fibre optic sensor (FOS) when it uses any of the fibre optic technologies. These sensors detect the brittle fracture, crack, temperature variation, corrosion, pull out and strain in the tendons. One of the main advantages of FOS is its ability to integrate the number of measurement points along the single fibre line. These sensors are considered to be the best option in many cases because of their characteristics: (i) immune to electromagnetic interferences; (ii) chemically inert; (iii) withstand high temperatures; (iv) small and light weight; (v) free from corrosion; and (vi) excellent transmission capability.

FOS is mainly used to measure the strain and temperature variation of the objects. Fibre Bragg grating (FBG) sensor is a commonly used type of FOS, which works on the principle of optical reflectance. Some of the FOS that are commonly used in SHM are (i) FBG; (ii) Fabry–Perot interferometric sensors; (iii) distributed Brillouin and Raman scattering sensors; and (iv) SOFO interferometric sensors. FOS is successful in crack monitoring. All the segments of the fibre act as a sensor in distributed sensing and it is the best option for monitoring large structures. The working principle of these sensors is based on the modulation of the light intensity transmitted through the fibre. The Brillouin FOS technology features distributed sensing and it can simultaneously measure the value of the strain, and also exactly locate the strained point along the structure where the sensor line is embedded. These distributed sensors do not require a prior knowledge of the crack location. The FOS has the capability to integrate the measurement point along the single fibre line. Therefore, a single fibre line can detect several cracks. The Brillouin FOS was implemented to monitor a bridge A6358 in Miller county, MO, USA and the load test was carried out by positioning six calibrated trucks along the bridge simulating different load conditions. One of the main applications of distributed fibre optic sensing is pipeline crack monitoring.

(c) *Strain monitoring using FOS*

The strain measurement is generally done using FBG sensors and Fabry–Perot sensors. The FBG sensors are embedded into the concrete structure during the manufacturing or fabrication process. These types of sensors are embedded into the structure and are used to measure the strain value in many lab scale models and real-time practices. To detect and locate the strain variation throughout the whole product, the composites are embedded with the long gauge FBG during the manufacturing process by vacuum-assisted resin transfer moulding. The optical domain reflectometry is used as the measurement system to measure the distribution of temperature and strain along the FBG. FBG sensors, monitoring high frequency dynamic strains, are used to detect the impact on the metal plate. The FBG sensor pairs are used such that at least one of the sensors will be able to pick up the stress wave. Time frequency analysis of the ultrasonic waves propagating through multiple FBG sensors is used to identify the location of the stress.

(d) *Temperature monitoring using FOS*

The FBG sensors are used to measure the temperature variations. The dual sensitivity of the FOS to the temperature and strain is a serious limitation. The temperature variation leads to abnormal

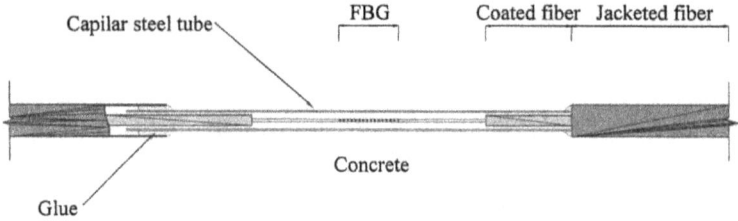

Figure 4.6. Fibre optic temperature sensor.

strain values and this is a serious issue when measuring strain and temperature values simultaneously. This can be overcome by using reference grating in thermal contact with the structure and this will not respond to the local strain variation.

Figure 4.6 shows the simple temperature sensor based on FBG technology. These sensors can be easily embedded in the structure and the number of measurement points can be minimised by using the same sensor for strain and temperature measurement.

(e) *Corrosion monitoring using FOS*

FOS is used to monitor the corrosion of steel bars in concrete structures. The sensors are based on Brillouin and Bragg grating technology, the latter is used for strain and temperature measurement as well. When the corrosion occurs, the volume of the steel reinforcing bar expands increasingly due to the rust accumulation on the surface. The fibre Brillouin optical time-domain analysis (BOTDA) is based on this principle and the fibre coil winds around the steel bar; when there is an expansion in the steel bar, the coil is stretched and the change in tension strain can be measured using a BOTDA analyser. Figure 4.7 shows the commonly used corrosion expansion sensor.

4.7.2.   *Future SHM technology*

(a) Carbon nanotube sensor

The smart materials developed with nanotechnology can measure physical properties in a more advanced way than the current

Figure 4.7. Brillouin corrosion expansion sensor.

technologies. Carbon nanotubes (CNTs) are the most used smart nanoscale materials because of their physical properties. These materials are extremely strong and have high mechanical and piezoresistive properties. The mechanical stiffness is very high approximately about 1 TPa. When there is a deformation, these CNT fibres will change the conductance of the material. With 1% of single-walled nanotube (SWNT), the polymer composite will become a self-sensing structure. Even though CNTs enhance the mechanical property of the polymer composite structure, it has some dispersion and adhesion issues during manufacturing.

Fully integrated strain sensor is designed by fabricating single-walled nanotube poly(sodium 4-styrenesulfonate)/poly(vinyl alcohol) (SWNT-PSS/PVA) thin film system, which acts as a large area sensing skin on structure when connected to an electrical readout circuit. The integrated sensors were tested in a partial steel structure as in case of a real-time SHM application. The experiments were carried out by placing integrated CNT strain sensors and the traditional metal foil strain sensors in the web of the steel beam. The output of the strain sensors was interfaced to a Narada wireless sensor node for data collection. The cyclic four-point loading test is carried out and the results show that the performance of the sensors is with a high sensitivity in excess of $5V/\varepsilon$. The sensors were also tested and validated using a steel sub-assembly specimen loaded to failure. Impact tests were carried out in CNT-enhanced composites patterned with a silver ink electrode grid. The results show clear changes in the regions close to the grid lines in the impact zone. As a result of the impact damage, a full field representation of damage location is obtained from the collected data.

(b) Artificial neural network in SHM

Artificial Neural Networks are the biologically inspired computational model. It is a distributed processing system, which has many interconnected parallel processors or neurons that can learn and generalise. The learning process determines a relationship between input and output variables based on the pattern. Multi-layer perceptron is the most commonly used neural network for SHM. It has various layers dedicated for input signals, output and a hidden layer of neurons. These layers are interconnected, the system is trained with the desired output, and any variation between the desired and actual output is determined. Back-propagation neural (BPN) network is another method used in SHM applications. It also has several layers comprising input, output and hidden neuron layer. The neural network is trained with dataset to identify the difference between the frequency response spectra of damaged and undamaged structures. The damage detection by using BPN is experimented in a five-storey steel frame. In this, the BPN is trained with the dataset of the scaled down model. In another set of experiments, the damage detection of a 20-bay planar truss is performed by a three-layer BPN, trained with the frequency response function to detect the damage cases. The damage identification using probabilistic neural network (PNN) is experimented in a four-storey steel frame and the PNN is trained with four damage patterns. The performance of PNN is compared with that of the BPN. The PNN has better performance and is suggested as the best opted method for damage localisation.

Researchers proposed a one-dimensional convolutional neural network for structural damage detection. In this study, the raw data of vibration signals obtained from accelerometer are classified without any pre- or post-processing of the data. It also exhibits a superior damage detection and localisation accuracy. The neural network models designed from data are rapidly increasing in the field of damage detection. Building such models requires a complete understanding of the structural behaviour, obtained data, available neural network models and the selection of appropriate models. Artificial neural network is an emerging technology for the future smart structures provided it is applied appropriately.

4.8.   Components of SHM System

Hardware design of the SHM system is divided into four functional modules as follows:

- sensor unit;
- DAQ unit;
- computational core;
- wireless communication channel.

For each functional module, commercially available hardware components are chosen based on the analysis of technical details of each component, its versatility, availability and cost. Prime objective is to arrive at a layout of sensors costing lesser and functionally suitable. The performance of each functional module of SHM is described in the successive headings. Figure 4.8 shows the components of the proposed SHM system.

4.8.1.   *Sensing unit*

Sensor units are the fundamental building blocks of the SHM system. Typically, the sensor-based SHM systems are intended to

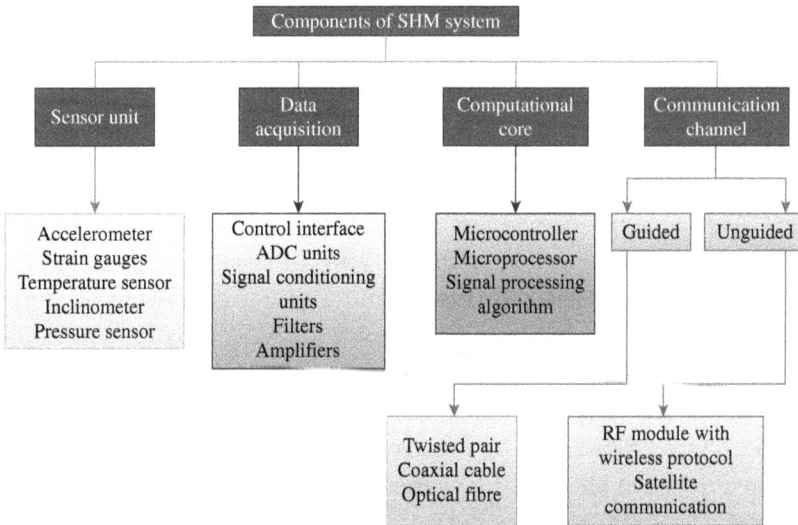

Figure 4.8. Components of SHM.

measure two quantities in general, kinematic quantities (for example, acceleration, inclination, strain and displacement) and environmental quantities (for example, temperature, humidity, wind, etc.). Wireless sensor units are proposed for monitoring the response of the offshore structures in this research. As offshore structures are massive and complex in nature, it imposes an additional requirement of performance level for the monitoring system. Considering the recent advancements in sensors, wireless sensing unit and smart sensing elements are chosen for the present requirements. In the selection of the type of sensor, the basic criteria considered are as follows:

   (i) capability to capture both local- and system-level responses;
  (ii) capability to acquire data in both consistent and retrievable manner for long-term data processing and analysis;
 (iii) adaptability to operational and environmental conditions.

Sensors are chosen based on the commercially available sensor units that best match the defined sensing performance requirements.

4.8.2. *Data acquisition unit*

Data acquisition (DAQ) system along with the sensor units measures the physical response of the model. It consists of hardware and a PC with the required software. Basic parameters considered while choosing the DAQ units are sampling rate, resolution, number of channels and type of transducers. In this research, DAQ of the wired SHM system is chosen as Spider8, supported by Catman Express software. Wireless modules are chosen such that they match with the DAQ capacity of that of the wired ones. Sensing interface is designed to be compatible with all types of sensors with a lower or a higher sample rate. Traditional SHM system, consisting of analog sensors, requires analog-to-digital converter (ADC) at the DAQ end. With recent advancements in sensor technologies, MEMS sensor modules, which are inbuilt with ADC, are used in the present design. Further, wired SHM may be involved with a large number of sensors; distance between the sensing element and the DAQ system may also be longer leading to data degradation that arises from the noise. Wireless DAQ systems are used, which save the installation time and cost of structural monitoring application while maintaining the compatibility.

4.8.3. *Computational core*

Computational core will be a microcontroller unit, which has algorithms to interrogate the measured data. Computational core is a combination of data storage and processing unit, which needs memory to store the measurement data and required software. It consists of at least 256 KB of RAM memory, so that it can store the data temporarily till the time it has been transmitted. The processor consists of expandable memory by SD cards, operating system, USB flash drives, interfaces for external devices, etc. It is programmed to perform customised operation, such as destined data collection, storage, processing and transmission as required by the SHM system.

4.8.4. *Wireless communication channel*

Design parameters that govern wireless communication channel in the current SHM design are: (i) data rate; (ii) open space range; (iii) encoding reliability; and (iv) radio frequency band a very few. With the recent advancements, wireless communication channel is designed based on required functionality, a form follows function approach. Main parameters of the communication channel are reliability and range. In case of wireless communication over a large area, possibilities of data loss due to interference, reflections and path loss may occur, which are vital design parameters addressed in the proposed WSN.

4.9. Conceptual Design of SHM System 1

Figure 4.9 shows the conceptual design of the SHM system 1 deployed in experimental investigations. Sensing and processing board consists of an accelerometer, microcontroller unit and a wireless transmitter.

The proposed SHM system 1 is designed for the preliminary data analysis to assess the vital requirements of health monitoring of offshore structures. It is important to note that the proposed design is customised for compliant structures where displacements are to

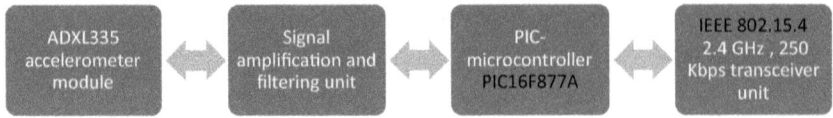

Figure 4.9. Conceptual design of SHM system 1.

monitored instead of member strength. As failure in compliant structures occurs mainly due to large displacements, the choice of sensors is made with this as a primary requirement. As miniaturisation and lightweight design are necessary for increasing the applicability, accelerometer ADXL335, which is based on MEMS technology, is used. The chosen accelerometer, ADXL335, is a tri-axial model and compact in shape, size and weight having signal-conditioned voltage outputs. It is a polysilicon, surface-micromachined sensor with signal conditioning circuitry, which is adaptable to open-loop acceleration measurement architecture. It is capable of measuring dynamic acceleration that results from shock or vibration and measures with a minimum full-scale range of $\pm 3g$. The scalded model of offshore compliant structure as attempted in the present research has lower natural frequencies; a bandwidth of about 50 Hz is seen to be sufficient for measuring its dynamic response.

Microcontroller unit used in the experimental investigations is a PIC16F778A microcontroller, which integrates a large storage memory and interface circuits. It includes built-in modules such as 8-bit high performance RISC CPU that features 256 bytes of EEPROM data memory, 10-bit ADC and 2 timers. The synchronous serial port, which is a part of the module, is configured as serial peripheral interface (SPI) or inter-integrated circuits (I^2C) bus and UART. PIC controllers are reprogrammed as they use the flash memory. The PIC16F778A microcontrollers are used to process the acquired signal from the accelerometer module, which is capable of withstanding high range of temperature variation. As the microcontroller is not compatible with the RS232 module, MAX232 is used to convert the TTL voltage level into RS232. Lead-Acid, rechargeable battery MR645 6 V, 4.5 Ah is connected to a 25 V-1000 μF capacitor, which in turn is connected to the transformer; power supply of 5 V input is regulated through IC 7805. Chip used on the board is well protected from the harsh environmental conditions by a hard cover. The microprocessing unit on the board gathers pre-processed analog and digital outputs from the sensors; processed data are then fetched to the wireless transmitting unit. As the data communication between the server and the base station is vital, wireless transceiver of 2.4 GHz is connected to the computing core. With the increase in distance between the sensor node and the server, power consumption also increases proportionately. Hence, wireless transmission in the present design is based on IEEE 802.15.4 ZigBee application. While it operates on 2.4 GHz frequency, wireless receiver unit is connected to a COM port of a PC through a Serial-USB converter.

4.10. Implementation of SHM System 1: Buoyant Leg Storage and Regasification Platform

Offshore platforms are of high strategic importance, whose preventive maintenance is top priority. Implementation of SHM to offshore platforms ensures safe operability and structural integrity. Prospective damages on the offshore platforms under rare events can be readily identified by deploying a dense array of sensors. Buoyant leg storage and regasification platform (BLSRP) is one of the most recent platforms to handle LNG storage and processing, which are highly hazardous. A novel scheme of deploying WSN is experimentally investigated on an offshore BLSRP, including postulated failure modes that arise from tether failure. Response of the scaled model under wave loads is acquired by both wired and wireless sensors to validate the proposed scheme. In order to validate the developed SHM system 1, the dynamic response of a scaled model of BLSRP is acquired using both wired and wireless sensor networks. Results are compared in both time and frequency domains to estimate the error of disagreement, if any. Wired sensors are connected to the DAQ system through wires and the data are further processed using central server, which is connected to the DAQ system. Wireless sensor nodes comprise of microcontroller unit and MEMS accelerometer. Acquired data are transmitted using the ZigBee module connected to the microcontroller. Central server receives the data transmitted by the Zigbee module.

Figure 4.10 shows the scaled model of BLSRP and the sensor board placed on the deck plate. Both wired and wireless sensors are placed on the scaled model to compare the acceleration of the deck acquired from both the nodes. Wired and wireless SHM systems are placed at the centre of the deck. Model is excited by a regular wave (wave height 10 cm) with varying periods from 1.2 to 2 s. DAQ of wired and wireless is carried out simultaneously. Table 4.1 shows the specifications of both wired and wireless accelerometer modules

Figure 4.10. BLSRP with wired and wireless SHM system 1.

Table 4.1. Specifications of accelerometer module.

Description	Wired	Wireless
Accelerometer type	B12/200 transducer	HBM ADXL335
Maximum range	±200 m/s^2	± 30 m/s^2
Sensitivity	80 mV/V	330 mV/g
Excitation voltage	1.8–3.6 V	1–6 V
Noise density	—	300 μg/$\sqrt{\text{Hz}}$

used in the study. Even though the sensor specifications are different, the operating frequency of the chosen sensors is within the desired range. Wired sensor has maximum range when compared to that of the wireless, but the compensation is done in terms of cost.

4.11. Preliminary Data Analysis

Accelerometer readings acquired through both the wired and wireless sensors are plotted in Figures 4.11 and 4.12, respectively. Preliminary analyses carried out on the acquired data showed that the maximum value acquired by the wired sensor is 0.62716 m/s^2. With due consideration to time delay, the corresponding maximum value of the wireless sensor is found to be 0.654 m/s^2, while the overall maximum value is 0.817 m/s^2. Maximum value acquired from wireless sensor does not match with that of the wired one due to the noise ratio in the device and differences in the sensitivity of both the devices. Considering the overall maximum value of wired and wireless accelerometers, error is found to be about 30.12%. While considering time delay and comparing the corresponding peak at a specific time, it is seen that the difference is about 4.2% while the peak signal-to-noise ratio is about 9%.

PSD of the data acquired by both the type of sensors is also plotted. It is seen that there exist few mismatches between both the PSDs. In case of the wired sensor, a distinct peak is seen at 0.5 Hz and consecutive peaks of low magnitudes are recorded at 1, 1.5 and 2.5 Hz. But the peaks in the wireless sensor acquisition are not explicit; frequency component occurs at multiple frequencies due to the presence of random noise. This difference in behaviour is attributed to many factors: (i) delay caused in the transition time; (ii) white noise in the wireless sensor data; (iii) sensitivity; and (iv) type of measuring methods. It is observed that random noise and presence of outliers are the main factors to be considered in post-processing. It is interesting to note that the above differences essentially arise due to the characteristics and not due to the system architecture adopted for the acquisition. As observed, the

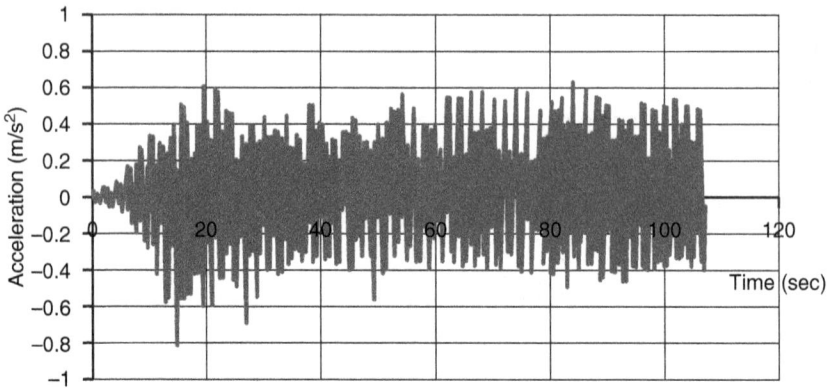

(a)

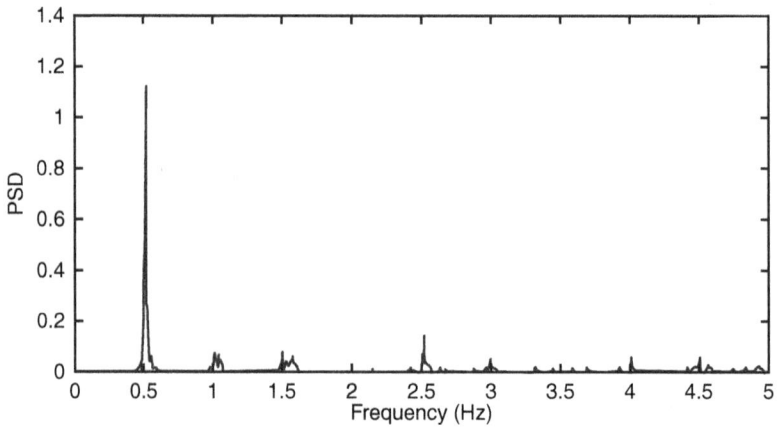

(b)

Figure 4.11. Surge response of deck using wireless sensors: (a) time history; and (b) PSD.

qualitative value of the acquired data does not change significantly. This strengthens the adopted architecture of WSN for health monitoring of offshore platform used in the current study. Improvements are not only made in the sensor DAQ but also with the processor units and data-transmitting techniques.

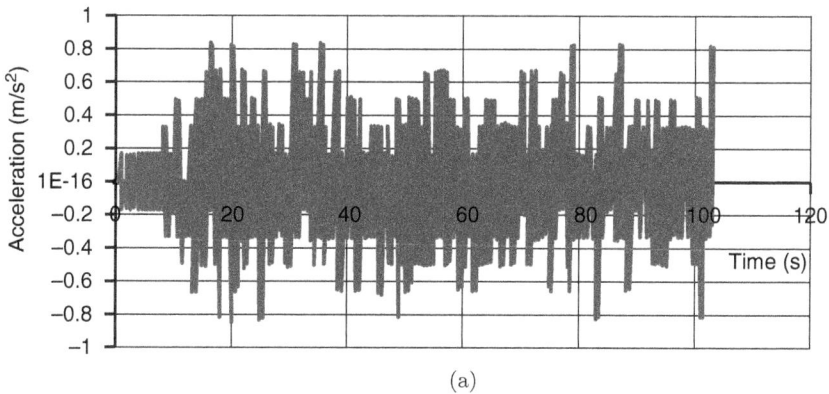

(a)

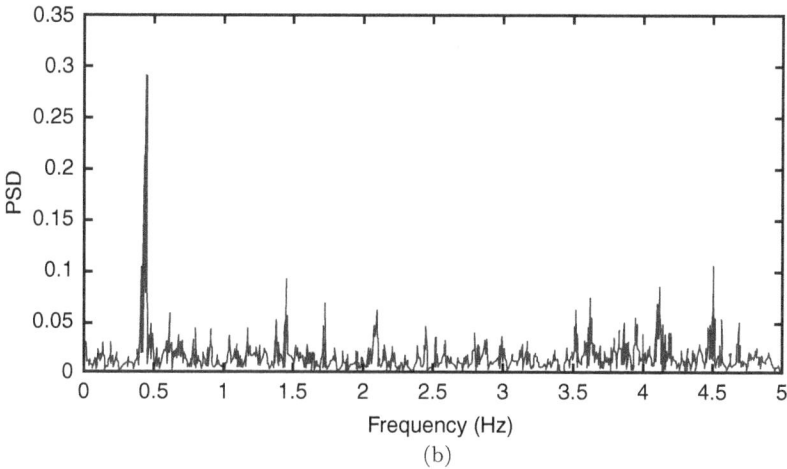

(b)

Figure 4.12. Surge response of deck using wireless sensors: (a) time history; and (b) PSD.

4.12. Observations and Improvements on SHM System 1

Preliminary analysis is carried out on the acquired data using SHM system 1 to upgrade the design. The sensor used in this study is a combination of accelerometer and gyroscope, which can measure both translational and rotational responses. As discussed in Section 4.11, it is seen that the PSD of the wireless data shows a higher percentage of noise. Observed difference between the PSD of wired and wireless sensors, at any time interval exceeds about 20%. This may be attributed to the design of sensor nodes and microcontroller unit of the design. Primary difficulty arises due to the fact that SHM system 1 is not capable of processing the data but can only acquire and transmit the data. To overcome this difficulty, a new and improved design of SHM system 2 is proposed in which the processing unit will be an integral part of the design itself. Sensor nodes shall require a separate ZigBee module, which is connected through the RS232 and a hyperterminal tool is used to collect the data in parallel. Delay in data transmission that may cause signal loss is addressed by choosing advanced transmission protocols. It is decided that sensor nodes should communicate directly with the central server, which can reduce signal-to-noise ratio and delay in transmission. Further, the new system is also enhanced with an automated AMS and a user interface.

4.13. Improvement in Sensor Units

Sensor unit refers to the sensors and their interfacing units for input signals. The study is mainly focused on the compliant offshore structures, which are designed to undergo large displacements. Stresses in members will be limited but the displacements are intentionally kept large. In compliant offshore structures, displacements are required to be measured and not the stresses in members. Under the action of environmental and operational loads, the offshore compliant structures will undergo both translational and rotational motions. The major parameters studied in the current research are displacements and rotations (not strain and loads). Thus, the sensor units to measure acceleration and displacement are chosen based on their functional requirements. Sensor units are chosen based on sensitivity and operating frequencies apart from ensuring their compatibility with that of the processing units. Scalability of the chosen sensor modules is also considered while selecting the appropriate hardware. Based on the above factors, the sensor unit opted for this study is MPU6050, consisting of MEMS-based accelerometer and gyroscope module. It comprises tri-axial accelerometer, tri-axial gyroscope and digital motion processor. It also features three numbers of 16-bit digital ADCs for digitising the output received from accelerometers and inclinometers. In the accelerometer module, each axis has separate proof mass while the displacement along each axis on the corresponding proof mass computes the acceleration in the respective axes. Sensor modules are chosen with two additional characteristics: (i) reduced settling effects; and (ii) control of sensor drift by elimination of board-level cross-axis alignment errors between the sensors. The combined operating current of sensors is limited to about 3.8 mA (full power with accelerometer operating at 1 kHz sample rate). For precision tracking of both fast and slow accelerations and displacements (both rotational and translational), sensor modules also feature a user-programmable, full-scale range of $\pm$ 250, $\pm$ 500, $\pm$ 1000 and $\pm$ 2000°/s (dps) for the inclinometer and full-scale range of $\pm$ 2g, $\pm$ 4g, $\pm$ 8g and $\pm$ 16g for the accelerometer. Full-scale range of $\pm$ 2g is opted for this study with 16,382 LSB/g sensitivity of the device. Accessibility for internal registers is enabled by deploying I²C at 400 kHz; an alternate could be to use SPI at 1 MHz. The reason for choosing the former is that it has a two-wire interface comprised of signal serial data (SDA) and the serial clock (SCL). As these two lines are bidirectional, the sensor module can either act as a master or a slave. Generally, MPU6050 acts as a slave but while communicating with the processor, it acts as a master node. Connected to the processor board through general purpose input output (GPIO) pins, sensor nodes and processing unit shall acquire the data and also transmit to the base station in parallel. The MPU6050 pin description is given in Table 4.2.

The AFS_SEL is a 2-bit, unsigned value of accelerometer configuration at the hex address $0 \times 1C$, which selects the full-scale range of acceleration. Similarly, Gyroscope configuration pin, allocated at the hex address $0 \times 1B$, is set to trigger the gyroscope self-test and also to configure for the full-scale range. The $0 \times 6B$ is the PWR_MGMT register, which configures the device power mode and clock source.

Table 4.2. MPU6050 pin description.

No.	Name	Description
8	VLOGIC	Digital I/O supply voltage. VLOGIC must be $\leq$VDD at all times.
9	AD0	I^2C slave address LSB
23	SCL	I^2C SCL
24	SDA	I^2C serial data

While the sixth bit of this register will reset the device when set to 1, the power mode can either be set to sleep or cycle mode. Device switches between the sleep and wake up modes are set to acquire a single sample at a rate specified by the LP_ WAKE_CTRL register. I^2C_ MST_CTRL configures the I^2C bus for multi-master control and 8 MHz internal clock. These registers are initialised as the first step in configuration using an interface code, which has been indigenously developed for the SHM system 2; pin configuration for initialising the device is executed using the following code:

$$\text{self.bus.write\_byte\_data}(0 \times 68, 0 \times 6b, 0 \times 01)$$

$$\text{self.bus.write\_byte\_data}(0 \times 68, 0 \times 1B, 0 \times 00)$$

$$\text{self.bus.write\_byte\_data}(0 \times 68, 0 \times 1c, 0 \times 00)$$

Offset value is calculated for each axis before the actual acquisition to adjust the error due to offset and drift value. In the MPU6050 module, there are three numbers of 16-bit ADC units to read the accelerometer output in all the three axes. For example, to read the X-axis output, ACCEL_XOUT_H and ACCEL_XOUT_L are the two registers, which have 8 bits each. ACCEL_XOUT has the 16-bit 2's complement value. Data within the internal register are continuously updated at the desired sample rate, which ensures that the registers will read the measurements from the same sampling instant. On its failure, the user shall be able to check whether it follows the single-byte read corresponding to the single sampling instant. All required library files are included in order to read and write the register values. 16-bit value is read first and the complement of the value is considered for data processing; the same procedure is carried out for all the axes. The code used for reading the abscissa values from the ACCEL_XOUT register is as follows:

```
high = self.bus.read_byte_data(0 × 68, 0 × 3b)
low = self.bus.read_byte_data(0 × 68, 0 × 3c)
val = (high << 8) |low
mask = (2 ** bitlength)−1
if val & (1 << (bitlength−1)):
     xval = val |~mask
else:
     xval = val & mask
```

Two major factors considered while dealing with the sensor measurements are noise reduction and number of outliers. Both noise and outlier presence will degrade the validity of the results. The source of noise in signal measurements may arise due to many factors: (i) presence of thermal noise; (ii) electromagnetic interference; (iii) sensor oscillation; and (iv) quantification of noise. In the proposed design of SHM system 2, efforts are taken to minimise the noise that arises from the above sources. This is also checked by validating the acquired data with that of the wired sensors. The effect of random noise, found from the PSD at 10 Hz, is about 400 $\mu g/\sqrt{Hz}$. Outliers are obtained by examining the spikes that appear in the RMS level of the data for the specified sample interval. Time series of these spikes are checked for any irrational measurements and subsequently filtered before processing the data. For the accelerometer module, the maximum output data rate is about 1 kHz, which is found to be satisfactory for the required experimental investigations.

4.14. Measurement of Rotational Responses

In general, MEMS accelerometer module will measure the linear acceleration and the local gravitational field. In the absence of this field, the output will be the value of the rotated gravitational field vector, which can be used to find pitch and roll responses. Between the two values of static acceleration and acceleration due to gravity, an inclined angle is generated. This angle corresponds to the tilt value of the sensor. The raw value of the accelerometer is read and the corresponding g value for the respective axis is calculated. Pitch and roll estimates are given by:

$$\text{Pitch} = \arctan\left(\frac{G_y}{\sqrt{(G_x^2 + G_z^2)}}\right), \tag{4.1}$$

$$\text{Roll} = \arctan\left(\frac{-G_x}{G_z}\right), \tag{4.2}$$

where G_x, G_y, and G_z are the linear acceleration in x-, y- and z-axes, respectively. The denominator of the pitch value is defined to be positive and thus Eq. (4.1) shall provide the values within the full range $[-90, +90]$. For the roll, the range is $[-180, +180]$. When the values of G_x and G_z are zero, then the roll becomes undefined.

4.15.   Processing Unit

Processing unit used in this study is the Raspberry pi ARMv7 processor. This is a low-cost computing device, which has an ARMv7 processor, random access memory and various other interfaces that are required to connect to the external devices. This unit needs a keyboard for command entry and a display unit to function as a stand alone device. But, when used as a web server, these peripherals are not required. Thus, pi-board acts as the perfect processor device without any peripherals and can be remotely connected to a central server unit. Pi-board has a 512-MB RAM as the internal programmable memory and the secure digital (SD) flash memory is configured in the processor board. To enhance the memory of the processing unit, an extended memory unit of 16 GB is added. This will enable the sensor node to store the collected data temporarily till the time it has been transmitted. Wi-Fi adapter is connected through a USB dongle with the help of which the pi-board is used for creating *ad hoc* networks and to connect to a wireless network; this runs on the 802.11n standard. As a computing device, it uses the Raspbian Wheezy OS as its operating system, which is a Linux-based open source OS. The connection between the sensor unit and processor board is established through the GPIO pins. The chosen Model B+

Raspberry pi-board has 40 GPIO pins, which includes the supply and the ground pins. The GPIO pins can accept input and output commands and thus can be programmed on the Raspberry pi. These GPIO pins read the connected sensor unit and are also used as an interface between the embedded protocols, such as I^2C and SPI. The processor unit deploys I^2C and serial peripheral interface protocols to communicate with the external device.

Sampling rate of the sensors is selected such that the device is able to detect changes associated with the maximum signal frequency. The signal can be reconstructed if the sampling rate is twice the maximum frequency component in the signal. The maximum signal frequency should be much lower than the Nyquist rate F_N in order to avoid over sampling and to ensure signal reconstruction. The sampling rate for this study is 40 Hz and the maximum signal frequency is expected to be below 10 Hz which is much lower than the F_N. Power optimisation is one of the most challenging factors in real-time health monitoring. The processor unit has four different power modes: (i) run mode in which all the functionalities of the core processor are powered up; (ii) standby mode in which the processor can be quickly woken up by the interrupts, till such time the core clocks are shut down and the power circuits are still active; (iii) shutdown mode; and (iv) dormant mode in which the core is powered down and the caches are on. In order to reduce the power consumption, standby mode is opted in the design. Power supply that is required to operate sensor nodes are sourced from the mobile power banks for the extended period of time during experimental investigations. Alternatively, in real-time monitoring, it is proposed to use alternate sources: (i) solar charger; (ii) special wall warts with USB ports; and (iii) alkaline batteries or rechargeable AA batteries with voltage regulator. Extending the lifetime of the SHM systems includes energy harvesting using solar, wind and vibration energy (Jahangiri *et al.*, 2016). In the recent past, sensor components with ultra-power circuit boards operating on nanowatts power are recommended for future generations (Lee *et al.*, 2016). Alternatively, power consumption can also be reduced by decreasing the number of samples to be acquired by the sensors. Data rate is reduced to a minimum thereby reducing the power consumption by saving the energy utilised for DAQ and storage. A reduction of about 80% of the number of samples in the data without affecting the accuracy of the data sequence shall result in energy saving of the sensors (Allipi *et al.*, 2010).

4.16. Wireless Communication Channel

Wireless DAQ systems save time and money in a variety of structural monitoring and testing applications. Several wireless communication protocols that are readily available in the commercial industry are generally used for networking sensors for SHM. Some of the commonly used protocols are IEEE 802.11 including Wi-Fi ranging up to 10 km with Yagi antennas operating under ideal condition. Alternatively, IEEE 802.15.2 protocols are used as their power consumption is lower. Further, in case of wireless sensors, IEEE 802.15.4 protocol is used with high gain antennas but upto a range of about 300 m. In this study, IEEE 802.11 protocol is used in the operating frequency range of 2.4 GHz, enabling the data transmission between sensor nodes and the base station. Wi-Fi adapter connected to the processor board transmits the data from sensor nodes to the central server. There will not be any packet loss as the TCP layer will handle this issue in the data layer. The TCP is capable of detecting failed packets, if any, and further retransmits them automatically. There will not be any order mismatch in the received packets as the packets will be sorted based on the header information. This ensures the reliable delivery of data being transmitted using TCP. However, in real-time monitoring deployed in offshore platforms, transmitting in the chosen frequency range with the available standard protocol is a challenging issue. Technology for communication in the offshore platforms should support voice over IP, broadband data and video communication services for different topologies of sensors. Improvements in satellite, very small aperture terminals (VSATs) and antenna systems shall be useful to meet the demand for higher bandwidth on offshore platforms. VSAT services support primary business processes, high quality voice calls and broadband Internet services when used on remote locations like offshore platforms.

4.17. Conceptual Design of SHM System 2

The conceptual design of the SHM system 2 is shown in Figure 4.13. A typical sensor node is a combination of sensing unit, processing unit and communication channel. Sensor nodes are designed to monitor the integrity of the structure, acquire and transmit the data to base station, while the server at the base station will further process the data. The SHM system is designed to remain in a sleep mode when the structure is idle. Even under a minimum displacement in the structure, sensors will become active and will start acquisition. Sensor nodes are placed at the optimised location; these locations are identified by carrying out numerous trial experiments with the objective of measuring the critical response. In this study, the position of sensors is chosen based on several trials conducted on the scaled models. However, previous researchers have also hinted about possible locations of three maximum amplitudes of responses, which are referred in the literature (Chandrasekaran and Lognath, 2017). Further, sensor nodes are not only chosen to be located at possible failure points, but also at sections that are expected to undergo maximum displacements even under normal load cases. The main objective is to monitor the entire structure as far as possible with

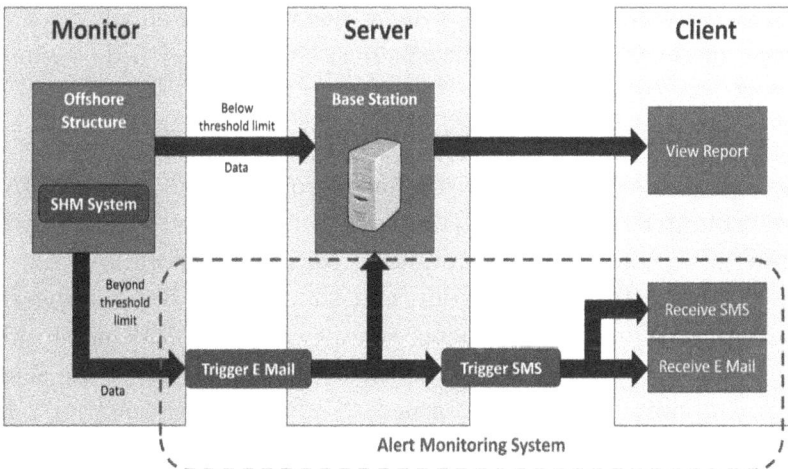

Figure 4.13. Conceptual design of SHM system 2.

the possible minimum number of sensor nodes. In this experimental study, sensor nodes and the base station are connected to a similar single network. There is no master slave node configuration as it is for a lab scale model and minimum number of sensor nodes is used for monitoring.

4.18.   Alert Monitoring System

Alert monitoring system (AMS) generates quick reliable reports when the acquired values exceed the preset threshold values. Threshold values are identified by benchmarking the average of previous records in non-critical conditions for the same structure; this value will determine the damage level detected by the sensor nodes. However, the SHM design has a provision to update or change these preset threshold values at the client interface. If the acquired data is below the threshold limit, it is considered that the structure is in healthy state. On exceedance of this value, the processing unit will trigger the email alert and transmit data to the base station. The server in the base station will process the data and perform a detailed analysis. Subsequently, an alert message will be displayed in the user interface. The alert message displayed in the monitor reduces the workload of the system administrator as complete data are not displayed. The server will further trigger an email and SMS to the registered mobile number associated with the SHM system.

4.19.   Calibration of SHM System 2

The dynamic behaviour of the accelerometer module is tested and calibrated using the shake table experiments with constant temperature under lab environment. Maximum effort is taken to free the test site from vibration and sound noise by isolating the test setup. Wired accelerometer module and the MPU6050 chip, along with the wireless sensor node, are placed on top of the shake table. Initially, tests are performed at constant frequency of 2 Hz at a displacement of 10 mm. Shake table experimental setup for calibration of SHM system 2 is shown in Figure 4.14. The shake table experiments are carried out for various frequencies ranging from 2 to

Figure 4.14. Shake table experiment setup.

5 Hz at various displacement levels of 10, 20 and 30 mm. Responses of both wired and wireless accelerometer modules are analysed in both time and frequency domains.

Plots of the acquired data are shown in Figures 4.15 and 4.16. It is seen from the figures that the frequency obtained from the wired SHM system is at 2.1 Hz while that of the wireless SHM system is at 1.95 Hz; the calculated error in the frequency is about 7.69%.

Another set of calibration test is also carried out at an increased frequency from 1 to 4 Hz. Time- and frequency-domain responses for this test are shown in Figures 4.17 and 4.18. It is seen from the figures that there is a gradual increase in the amplitude over a wide range of frequency, which is successfully captured by the proposed wireless SHM system. Signals obtained from both wired and wireless systems are not clipped at their peaks with the increase in the amplitude. The frequencies obtained by the two systems match closely with the maximum variation of about 10%. An exact match is not possible as the specifications of both sets of sensors are different within the same SHM layout. This calibration test ensures that the dynamic

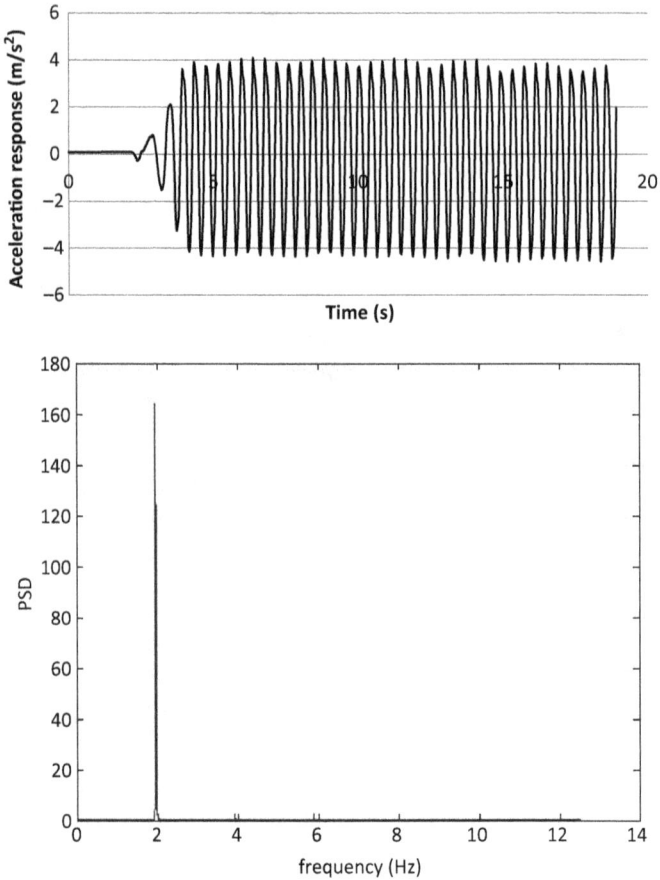

Figure 4.15. Acceleration time history and PSD of wired SHM system.

behaviour of the accelerometer module of SHM system is suitable for its implementation further in the experimental studies.

4.20.   Comparison of SHM Systems

Table 4.3 compares both the SHM system designs used in the preliminary studies. By observing the data in the table, it is seen that the wireless SHM system is superior to that of the wired system. It shows many advantages. However, among both the wireless systems, SHM system 2 shows distinct advantages over system 1.

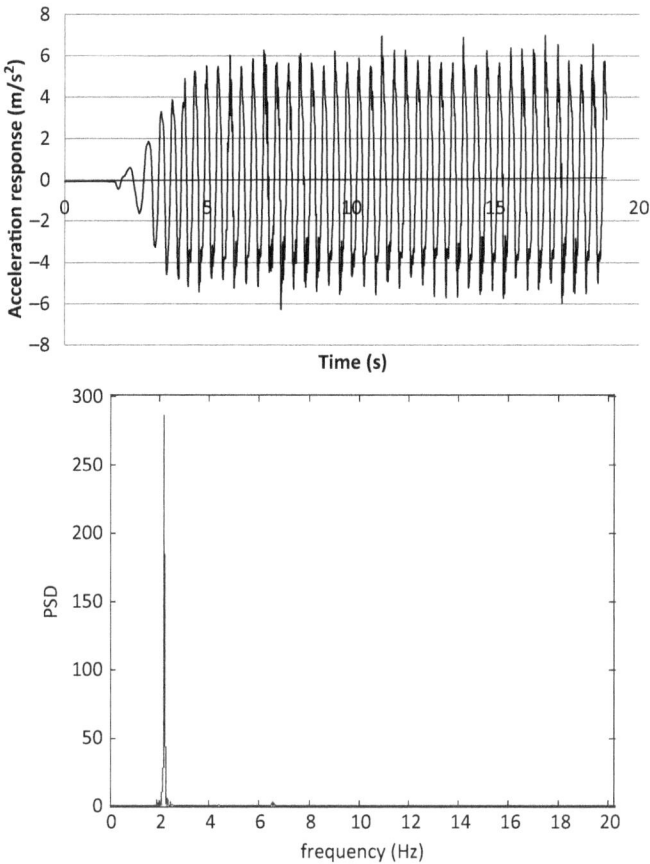

Figure 4.16. Acceleration time history and PSD of SHM system 2

Many drawbacks in system 1 are addressed in system 2. Hence, SHM system 2 is used in further experimental investigations.

4.21.   Challenges in Real-Time Implementation

Although the proposed SHM is successful in lab scale, there are a few factors, which need to be considered while implementing this design in real-time monitoring. Sensor network is developed based on the requirements of the experimental model of offshore compliant structures. In real time, the frequency range of the structures can be different and hence sensors need to be chosen based on the operating

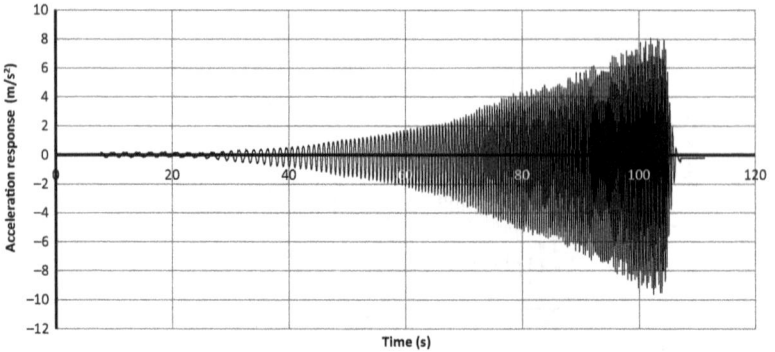

Figure 4.17. Acceleration acquired through SHM system 2.

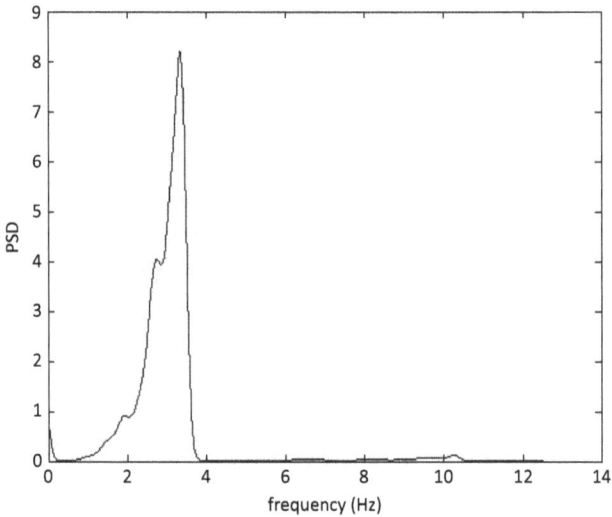

Figure 4.18. Frequency response of acceleration of SHM system 2.

frequency band. While layout of the sensor network and interfacing of hardware components remain unaltered for both the lab scale and real-time monitoring, bandwidth and latency issues are vital while implementing the software for the real-time monitoring. In addition, hardware configurations of the sensor units should be chosen based on the data requirements. For long-term monitoring, it is proposed that the sensors should be embedded in the structure while construction.

Table 4.3. Comparison of different SHM systems.

Wired	Wireless SHM system	
With 393B04 PCB module and Spider8 DAQ	With PIC microcontroller and ADXL335-802.15.4 protocol (SHM system 1)	With Raspberry pi and MPU6050-802.11.x protocol (SHM system 2)
Sensors are physically connected	Sensors are not physically connected	Sensors are not physically connected, but work as independent modules
Installation is complex and time consuming	Easier and quicker to setup	
DAQ unit will collect data from the sensor unit	Central server will collect the data from sensor nodes through 802.15.4 protocol and then make it visible in public domain	Local database will collect the data at sensor nodes itself and then transmit it through 802.11.x protocol to make it visible in public domain
Central server should be connected through wires to the sensor nodes	No wired connection is required. Central server should be placed in proximity to the acquisition node	Central server can be placed anywhere as the database will be uploaded directly to the web server
Data loss is lesser for lesser distance of layout	Probability of data loss is high in comparison with that of the wired network	Probability of data is very less as the data are stored in a local database in the system itself
There is no noise interference	Signal-to-noise ratio is seen to be significant	Noise interference is comparatively lesser

In real-time implementation, one of the main challenges of sensor nodes is to withstand human interventions and harsh environment. Further, IEEE 802.11 protocol used in lab scale is not compatible with all kinds of ocean environments; alternates include voice over IP, broadband data and video communication services for different topologies of sensors. Improvements in satellite, VSATs and antenna systems will enable it to meet the demand for higher bandwidth in offshore platforms.

4.22. Health Monitoring of Tension Leg Platforms in Lab Scale

Experimental investigations on a scaled model of tension leg platform (TLP) are carried out at 4m flume. Primary idea is to investigate the successful application of the proposed design of WSN, which is then subsequently applied to analyse postulated failure cases of the platform. An extension of AMS is also proposed to make useful application of the proposed WSN in terms of successful SHM. Scale of the models, choice of sensors to acquire data, and their integrated networking are indigenously designed as a part of this investigations (Thailammai, 2016).

SHM system includes choice of appropriate sensors for measuring displacements, load sensors for measuring dynamic tether tension variation and inclinometers to measure rotational displacement of members. Sensitivity and frequency range of sensors are chosen after considering the typical displacement values of scaled models and input frequencies of wave loads. However, chosen sensors may not be suitable for measurements on prototypes as both the range of frequency and amplitude of measurements may lies outside their range and capacity. Networking of sensors are designed only according to the chosen sensors but can be applied conceptual for prototype networking as well with appropriate modifications. Two sets of SHM systems are proposed — wired and wireless. Data acquisition is done on scaled model using both the systems. Results obtained by wireless are compared with that of the wired to validate its system architecture.

4.23. Components of Wired System

4.23.1. *Accelerometer*

Accelerometer used in the wired SHM system is a PCB 393B04 accelerometer module. PCB393B04 is an integrated circuit piezoelectric (ICP) sensor, which is used to measure acceleration time history over a definite period of time (T). It is also capable of measuring dynamic pressure, force and strain values. In this study, accelerometers are used to measure translational response (surge, sway and heave) of the scaled model of offshore platforms. Only compliant platforms are considered for the study due to their large displacement sensitivity. Accelerometer impediments the piezoelectric material as sensing element, which converts the physical strain value to electrical signal. The output of the accelerometer module is connected to a signal conditioner before it is connected to the DAQ channel. On excitation, force exerted by the mass on piezoelectric material generates electric output through the electrodes. For a constant mass, the force exerted by the piezoelectric material will be proportional to the external vibration. The output signal of $1\,\text{V}$ corresponds to $9.81\,\text{m/s}^2$ acceleration, as calibrated; the weight of the accelerometer module used in this study is about 48 g. Figure 4.19 shows the typical accelerometer used in the study. Specifications of the accelerometer module used in this study are given in Table 4.4.

Figure 4.19. Piezoelectric accelerometer.

Table 4.4. Specifications of accelerometer.

Description	Wired
Accelerometer	393B04
Type	ICP
Number of axis	1
Range	$\pm 5g$
Sensitivity	$1\,\mathrm{mV}/g$
Noise performance	$0.30\ \mu g/\sqrt{\mathrm{Hz}}$

Figure 4.20. Inclinometer.

4.23.2. *Inclinometer*

Inclinometer used in the wired SHM system is shown in Figure 4.20. *Posital fraba* is a biaxial inclinometer, which measures the inclination in two perpendicular axes. Specifications of the inclinometer are given in Table 4.5. For successful measurement, inclinometer requires an input voltage of 10 V. The angle of inclination is determined based on the conductivity measurements over a number of plane electrodes. The cell is filled with a conductive fluid at the bottom and electrodes are placed parallel to the axes. By applying an AC-voltage across the plane of electrodes, an electrical stray field is created. When the inclinometer is tilted, the level of conductive fluid over the various electrodes changes, resulting in change of conductivity. The stray field also changes, which measures the inclination. The output will

Table 4.5. Specifications of inclinometer.

Physical parameter	Value
Range	±80°
Axis	2
Supply voltage (V)	10–30 V DC
Interface	RS232 & Voltage
Accuracy	0.10°

be either in volts or in degrees based on the settings. The weight of the inclinometer used in the present study is about 50 g.

4.23.3. *Load cells*

Ring-type load cells are used to measure dynamic tether tension variation during the experiments. Figure 4.21 shows the typical load cell used in the current study. Passage of waves causes variable submergence on the partially submerged members of the platform, which causes imbalanced pre-tension in tethers. This variation, which is dynamic in nature, is captured using load cells. Load cells are fabricated using stainless steel ring of 32 mm diameter, 10 mm wide and 2 mm thick. Foil-type strain gauges are used for half-bridge configuration. Epoxy resins are used to protect the strain gauges from slashing water. The outputs of load cells are connected to the DAQ unit through a 15-pin connector. Load cells are calibrated with standard weights in the lab settings prior to the conduct of experimental investigations; load cell is calibrated.

4.23.4. *Data acquisition system*

HBM spider8 DAQ system is used for acquiring data of the desired parameters from the installed sensor modules, as described above. Spider8 DAQ system has eight channels for parallel measurement, which can be used to record non-electrical quantities such as inclination, strain, force, displacement, acceleration, pressure and temperature. In addition, it is useful to record electrical quantities, such as current, voltage, power and resistance. The DAQ system

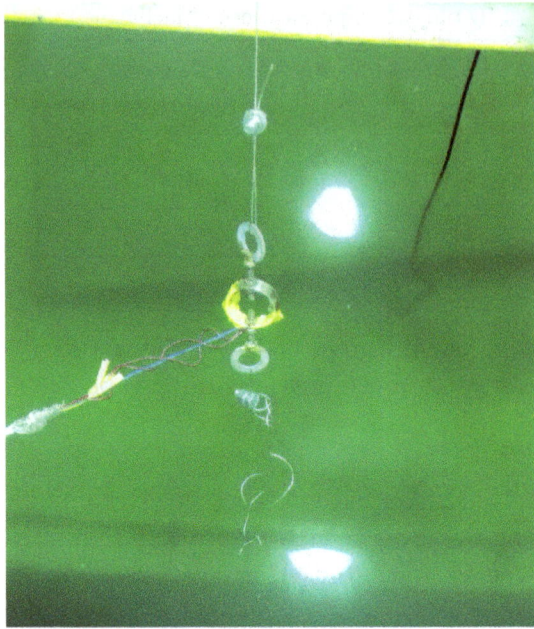

Figure 4.21. Ring-type load cell.

is interfaced with the personal computer through RS232 serial port connection. Catman software, pre-installed in the computer, is used to acquire the data in real time during conduct of experiments. Configurations of the sensors, including their scalability and calibration constant, are set in the software so that sensor devices do not have any variable elements. During experimental investigations, outputs from piezoelectric accelerometers, inclinometers and load cells are connected to the input channels of Spider8 through a 15 pin connector while the sampling rate of the DAQ system is set to 50 Hz.

4.24. Components of Wireless Networking System

The primary component of the wireless SHM system is the sensor node. Each sensor node is a combination of sensing unit, processor unit and the transceiver unit. The processor unit used is the Raspberry Pi board, which is a low-cost computing device. The processor of the Pi board is an ARMv7 processor with 700 MHz clock speed, which operates on a Linux-based, Raspbian Wheezy Operating System. While the Pi board integrates multiple input and output peripherals, the sensor unit is MPU6050 MEMS chip, which is connected to the processor through GPIO pins. To communicate with the external devices connected to these GPIO pins, either I^2C or SPI protocols can be used; the former is used in this investigation. Salient advantages of the chosen protocol are as follows: (i) it has sufficient memory and CPU cycles to program; and (ii) it controls and collects logs from the sensing units. An extended scalable device of 16 GB memory card is connected to the board to elevate the data storage capacity. This helps the storage of temporary data until they are transmitted further for post-processing. The Pi board is powered using an external battery for an extended period of time during the conduct of the experiment. Low power consumption of the chosen sensor nodes ensures possibility to operate from mobile power stations, which suits offshore platform installations. In case of a real-time situation, alternate solutions are applicable, such as (i) energy harvesting techniques — utilisation of solar, wind and vibration energy; and (ii) use of ultra-power circuit boards operating on nanowatt power for various WSN components (Lee *et al.*, 2016; Vahid *et al.*, 2016). The sensing unit used in this study is MPU6050, which is a combination of tri-axial gyroscope, tri-axial accelerometer and a digital motion processor. It is capable of measuring both the acceleration and rotation. It comprises 16-bit ADC to feature gyroscope and accelerometer output; latter are 3 in numbers. The sensing unit deployed in the network is capable of measuring $\pm 16g$, but a full-scale range of $\pm 2g$ with the corresponding sensitivity of 16,384 LSB/g is used during this study. The transmitter unit is connected to the Pi board, which acts as a transceiver unit. The sensor node acts as an independent module and sends data directly to the server through transmitter. Sensor nodes are implemented based on the four main steps — acquire, transmit, store and report.

Design of SHM architecture, deployed in the current research, is an indigenous attempt, which accelerates self-organisation of sensors.

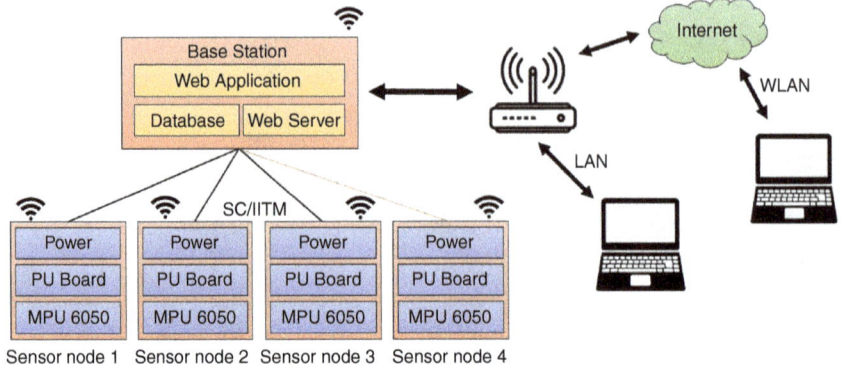

Figure 4.22. Wireless sensor networking architecture.

This avoids use of increased number of sensors as intermediate nodes will participate in forwarding the data packets between the source and the destination. Figure 4.22 highlights the wireless sensor networking architecture.

SHM architecture deployed in this study involves a component-level integration of multiple software and hardware units, which are required to develop a stable WSN. Four wireless sensor nodes are installed on the deck of TLP to capture its response and also to transmit the data through the transmitter. In this study, IEEE802.11 protocol is used, which enables the data transmission between sensor nodes and the central server. While all sensor nodes and the base station are connected to the same network, the server at the base station receives and stores transmitted data in a local MySQL database. A webpage is designed to access the collected data as a report and the web application is hosted through a static IP. This can be accessed globally by authenticated users.

Newly developed webpage is used as a reporting tool to read, analyse and generate alert messages when the sensor value exceeds the preset threshold limit of responses. In a real offshore platform, transmitting in this frequency is a challenging issue. Voice over IP, video communication service and broadband data for different topologies of sensors are some of the few technologies available to establish communication in the offshore platform. Improvements in satellite, VSATs and antenna systems will enable it to meet the demand for higher bandwidth on offshore platforms. The VSAT services support primary business processes, high quality voice calls and broadband internet services for offshore platforms.

Objectives of the developed wireless SHM system architecture are as follows: (i) to monitor the integrity of the structure; (ii) acquire the data; and (iii) transmit the data to the base station. The server at the base station will further process the received data. Unlike in wired systems, it is important to limit the transfer of data packets, which are not intended for monitoring the platform. On the other hand, if the platform motion is expected to remain well within the permissible limits of displacements in all active degrees of freedom, then the system architecture should enable an intelligent power-saving mode during which no redundant data is acquired and transmitted. During such period of time, the SHM system will be forced to remain in a sleep mode. Apart from saving power, it also reserves memory space for futuristic data on the server. Auto mode of the idle state will exit even for minimum displacement and will start acquisition. The threshold value of displacement is identified by benchmarking the average of previous records in non-critical conditions for the same structure. This value determines the damage level detected by the sensor nodes. However, based on the update or experience, threshold values can be changed by the end user at the client interface. If the acquired data is below the threshold limit, it is considered that the structure is in healthy state. When the value is beyond the threshold limit, the processing unit will trigger the email alert and transmit data to the base station. The server in the base station will process the data and perform a detailed analysis. It will generate alert messages, which will be displayed in the user interface, and will also trigger SMS to the registered mobile numbers associated with SHM system.

4.25. Experimental Model of TLP

In the present investigation, the SHM system is implemented in the scaled model of offshore TLP. They show rigid body motion of both deck and pontoons. TLPs are compliant-type offshore platforms generally used for oil exploration and production in deep waters. TLP has two distinct set of natural frequencies — very rigid frequency and very flexible frequency. TLPs are designed to exhibit high compliancy in horizontal plane and are very stiff in vertical plane. By virtue of design, surge, sway and yaw degrees of freedom have periods varying from 80 to 120 s (in prototype) while roll, pitch and heave degrees of freedom lie in the range of about 2–5 s (in prototype). The problem of health monitoring of such compliant offshore platforms is more challenging due to a wide range of distinct frequencies.

Offshore TLPs are compliant offshore structures, which experience large responses due to their compliancy. Tendons are deployed under pre-tension to counteract the excessive buoyancy. Response amplitude is observed to be more on its horizontal plane (Chandrasekaran *et al.*, 2016). Geometric sizing of members and the plan dimensions of the model are derived from the Auger TLP. The TLP model is scaled in 1:100 scale ratio using Froude's scaling. A view of the model under installed condition is shown in Figure 4.23. Four columns of the platform are fabricated using acrylic tubes of 250 mm outer diameter and 5 mm thickness; while the height of the column is 490 mm. Stiffeners are provided inside the column members to achieve the desired mass moment of inertia. Table 4.6 shows the geometric and hydrostatic parameters of the TLP model considered for the study.

Surge and heave accelerations are measured using piezoelectric-type accelerometers while the pitch response is measured using an inclinometer. Load cells are used to check the dynamic tether tension variations during the tests. Free vibration tests are carried out to determine the natural period and damping ratio of the scaled model in all active degrees of freedom. Initial displacements are given on the structure for surge free decay tests while a rotation angle is given for the pitch mode. Natural period and damping ratio are calculated using the conventional logarithmic decrement method. Based on the results, it is seen that the natural period of the structure in surge degree of freedom is about 9.65 s with a damping ratio is about 17%.

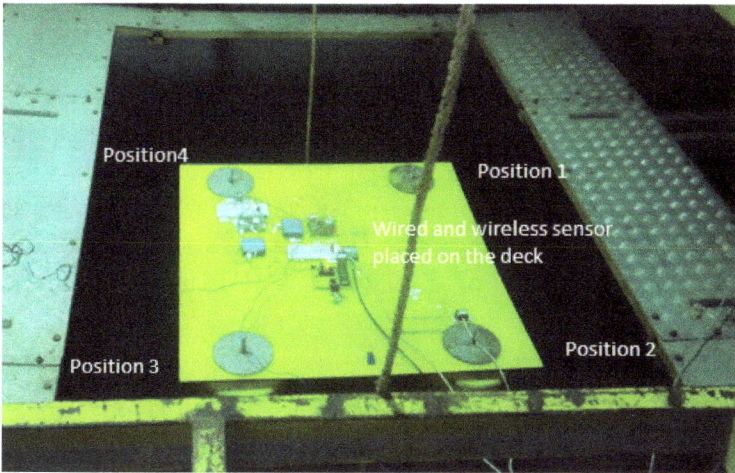

Figure 4.23. View of TLP model.

Table 4.6. Geometric and hydrostatic parameters of TLP.

Description	Prototype	Model 1:100
Length of the deck (m)	100.6	1.006
Width of the deck (m)	100.6	1.006
Column height (m)	48.5	0.485
Diameter of column (m)	25	0.250
Draft (m)	28.7	0.287
Width of pontoon (m)	10.7	0.107
Pontoon height (m)	8.5	0.085
Tethers diameter (m)	0.7	0.007
No. of tendons	12	4
Weight of structure (ton)	40000	40.0 kg
Centre of buoyancy (m)	12.4	0.124
Centre of gravity (m)	19.4	0.194
Metacentric height (m)	20.5	0.205

Natural period of pitch and heave are found to be 0.54 and 0.51 s with damping ratios of about 6.8% and 7.5%, respectively. Damping ratio varies linearly with the displacement. Dynamic response analyses are carried out under regular waves of 8–12 cm wave height and period ranging from 1.6 to 3.2 s.

4.26.   Postulated Failure Analyses

Experimental investigations are carried out for various cases to analyse the response of TLP as a part of the monitoring process. Postulated failure to cause damages on the scaled model is induced intentionally for experimentation purposes. Experiments are first carried out for an undamaged platform under regular waves to estimate the response characteristics. It is also useful to arrive at the threshold values of displacements. Postulated failure cases are not based on the past failures of real structures. The main focus is to assess the effectiveness of AMS. Postulated failure cases are introduced with limited amplitudes so that they do not cause any permanent damage to the model, which will prevent further investigations.

The extreme load conditions that occur commonly in the offshore structures were considered to develop the postulated failure cases. The stability of the platform may be affected by sudden impact loads on the deck of the platform and the tether pull out is the important condition that has to be considered in case compliant platform positions are restrained by tethers. Thus, the postulated failure is created by placing eccentric loads on the topside. Different postulated failure cases are as follows: (i) postulated failure case 1 with eccentric load over column 2 (wave approach direction); (ii) postulated failure case 2 with eccentric load over column 4; (iii) postulated failure case 3 with removal of tethers at column 2; (iv) postulated failure case 4 with removal tethers at column position 4 in addition to that of removal of tethers at leg position 3; and (v) postulated failure case 5 where the model is excited under 16 cm wave height, which corresponds to extreme sea state. Figure 4.24 shows the view of the platform under postulated failure case 1. Steel plates are added to increase the topside load, as shown in the figure. Increase in mass is about 18% of that of the total mass of the platform. Eccentric loading resulted in dynamic tether tension variation in tendons connecting column 2, which is essentially caused by change in the draft level. Other cases are postulated in a similar manner. Figure 4.25 shows the view of platform under postulated failure case 2

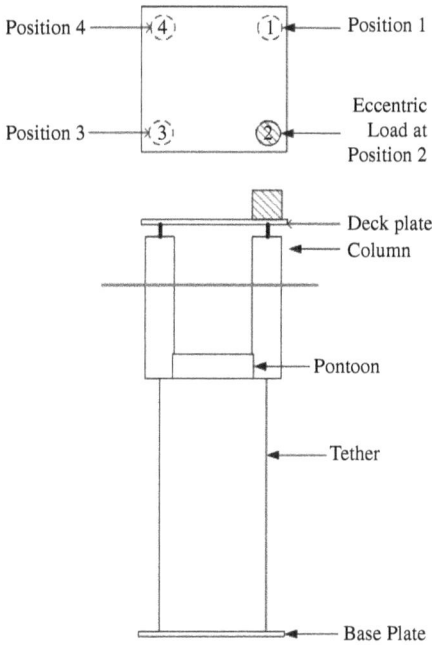

Figure 4.24. View under postulated failure case 1.

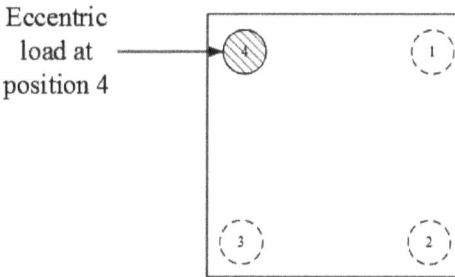

Figure 4.25. View under postulated failure case 2.

and Figure 4.26 shows the view of platform under tether removal conditions, which resembles postulated failure case 3.

Monitoring system in this discussion is concerned with that of the acceleration response of the platform; for exceedance of acquired acceleration in comparison with that of the undamaged case, it is

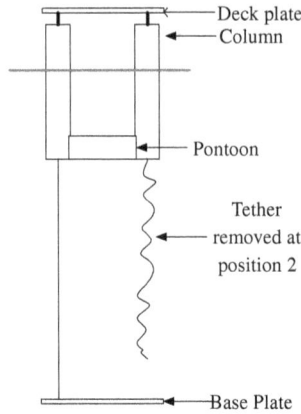

Figure 4.26. View under postulated failure case 3.

designed to raise an alarm as part of SHM. Four sensors are deployed, whose positions are changed for each set of the experiments. In every postulated failure case, it is assumed that the assumed failure alone occurs in the platform. Hence, the position of the sensors is changed accordingly. Initially, trial runs are carried out based on which the optimised position of sensor nodes is fixed. Sensor positions are chosen such that the response is the maximum and damage sensitive features are dominant. While sensor nodes are placed at the mass centre of the deck during trial runs, their positions are appropriately changed during each set of experiments based on the postulated failure cases. For example, during the experimental investigations on the TLP model, sensor nodes are fixed at the mass centre of the deck and also at positions marked as 1, 2, 3 and 4 in the figure. Placement of sensors in these positions differs based on the postulated failure cases and on the concentration of responses in that particular position.

4.27. Data Processing Techniques

Signal-based data analysis method involves processing the significant variations, in terms of the features of acquired time history of the data; alternatively it can also be on their corresponding frequency spectrum. This is generally carried out through various signal-processing techniques and algorithms. Signal-based damage detection method is further classified as: (i) feature extraction; and (ii) pattern recognition. Feature extraction method processes the time history of the response to extract the damage-sensitive features. When dealing with a large amount of data from multiple sensors, this process condenses the data into a smaller dataset, which can be better processed using statistical tools. Pattern recognition is a process of implementing algorithms on the extracted features to identify the damage state.

4.27.1. *Tilt sensing using tri-axial accelerometer*

Tilt angle is derived by measuring orientation in earth's gravitational field. The basic assumption is that the 3-axis accelerometer module is mounted on the structure oriented in earth's gravitational field g and undergoing linear acceleration (a), the output will be G is given by

$$G = \begin{pmatrix} G_x \\ G_y \\ G_z \end{pmatrix} = R(g - a)$$

where R is the rotation matrix describing the orientation of the module relative to earth's coordinate frame. Figure 4.27 shows the rotation axis and the corresponding coordinate system.

To solve the rotation matrix, it is assumed that the accelerometer has no linear acceleration $a = 0$; initial orientation is flat with earth's gravitational field aligned with the z-axis. Hence, the accelerometer output G is given by

$$G = \begin{pmatrix} G_x \\ G_y \\ G_z \end{pmatrix} = Rg = R \begin{pmatrix} 0 \\ 0 \\ 1 \end{pmatrix}$$

4.27.1.1. *Pitch and roll estimation*

Roll, pitch and yaw rotation matrices, which transform a vector under a rotation of the coordinate system in angles θ_1 in roll, θ_2 in pitch and θ_3 in yaw about the x-, y- and z-axes, respectively, are

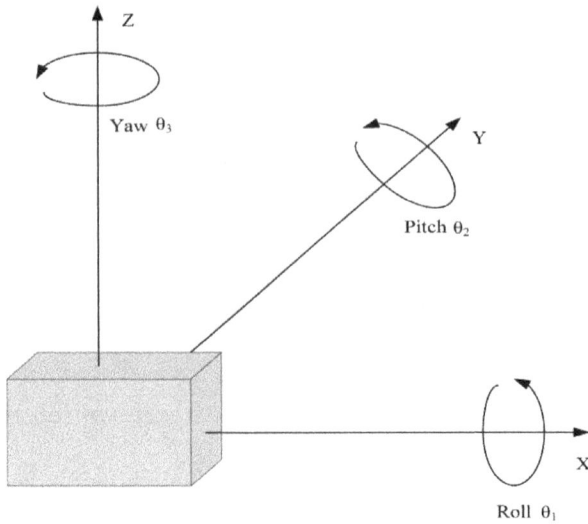

Figure 4.27. Rotation axis and coordinate system.

given by

$$R_x(\theta_1) = \begin{bmatrix} 1 & 0 & 0 \\ 0 & \cos\theta_1 & \sin\theta_1 \\ 0 & -\sin\theta_1 & \cos\theta_1 \end{bmatrix},$$

$$R_y(\theta_2) = \begin{bmatrix} \cos\theta_2 & 0 & -\sin\theta_2 \\ 0 & 1 & 0 \\ \sin\theta_2 & 0 & \cos\theta_2 \end{bmatrix},$$

$$R_z(\theta_3) = \begin{bmatrix} \cos\theta_3 & \sin\theta_3 & 0 \\ -\sin\theta_3 & \cos\theta_3 & 0 \\ 0 & 0 & 1 \end{bmatrix}.$$

One possible matrix is R_{yxz}, which is computed as follows and its effect on earth's gravitational field of $1g$ is determined:

$$R_{yxz} \begin{pmatrix} 0 \\ 0 \\ 1 \end{pmatrix} = R_y(\theta_2) R_x(\theta_1) R_z(\theta_3) \begin{pmatrix} 0 \\ 0 \\ 1 \end{pmatrix}$$

$$= \begin{bmatrix} -\sin\theta_2 \cos\theta_1 \\ \sin\theta_1 \\ \cos\theta_2 \cos\theta_1 \end{bmatrix}. \tag{4.3}$$

The sequence R_{yxz} is used to eliminate the yaw rotation θ_3. This allows solution for the roll θ_1 and pitch θ_2 angles. Equation is rewritten in relationship to roll and pitch angles and normalised with respect to G values as follows:

$$\frac{G}{\|G\|} = \begin{bmatrix} -\sin\theta_2 \cos\theta_1 \\ \sin\theta_1 \\ \cos\theta_2 \cos\theta_1 \end{bmatrix}$$

$$\Rightarrow \frac{1}{\sqrt{G_x^2 + G_y^2 + G_z^2}} \begin{bmatrix} G_x \\ G_y \\ G_z \end{bmatrix} = \begin{bmatrix} -\sin\theta_2 \cos\theta_1 \\ \sin\theta_1 \\ \cos\theta_2 \cos\theta_1 \end{bmatrix}.$$

Solving for the roll and pitch angles from the above, angles are computed according to the rotation sequence R_{yxz} and are given by

$$\tan\theta_2 = \frac{G_y}{\sqrt{G_x^2 + G_y^2}}, \tag{4.4}$$

$$\tan\theta_1 = \begin{bmatrix} \dfrac{-G_x}{G_z} \end{bmatrix}. \tag{4.5}$$

The above equations show results for roll and pitch angles. This is calculated by assuming that the rotation matrices do not commute.

4.27.2. *Frequency-domain analyses*

Frequency-domain technique is useful to analyse the stationary event, which is localised in the time domain. Some of the Fourier-domain methods are Fourier response spectrum, Fourier analysis, spectral analysis, Cepstrum analysis, etc. In this study, fast Fourier transform (FFT), PSD, and short-time Fourier transform (STFT) techniques are used to analyse the data in frequency domain. FFT is one of the best tools to identify frequency components present in the signal. It decomposes time-domain signal in terms of a set of basic functions. FFT assumes that the observed signal is infinitely long and periodic; period is equal to the record length of the observed signal. FFT reduces the number of operations to the order of $N \log(N)$. After computing FFT, the PSD function is estimated.

Let $X(t)$ be a time-varying function, which represents the acceleration time history that is acquired during the experimental investigations. Fourier transform of $X(t)$ is given by

$$X(F) = \int_{-\infty}^{\infty} X(t)e^{-j2\pi Ft}dt.$$

Fourier transform (FT) decomposes the signal into weighted combinations of sinusoids with different frequency. Transform finds the amplitude and phase of these sinusoids. For a specific value of F, the signal is correlated with the basis function $e^{-j2\pi Ft}$. Value of F ranges from $-\infty$ to $+\infty$. Complex correlation coefficient obtained for this value of $2\pi F$ is the Fourier transform coefficient. PSD of the signal represents the distribution of power across the frequencies composing that signal and is given by

$$S_x(f) = \lim_{T \to \infty} E\left\{ \frac{1}{2T} |X(F)| \right\} = \lim_{T \to \infty} E\left\{ \frac{1}{2T} \left| \int_{-T}^{T} X(t)e^{-j2\pi Ft}dt \right| \right\}.$$

The above equation can be interpreted as the expectation of FT of the signal computed over an infinite period. In case of Fourier transform, only global features of the signal are extracted in the frequency axis as there is no localisation of the features across the time axis. This is seen as one of the major deficits of FFT. Transform is the result of summation across the entire length of the signal. Hence, it is stated as a method that has a good frequency-resolution but poor time-resolution. In case of the SHM system, FFT can identify the damage by the presence of frequency spikes, but this damage identification is purely based on the information extracted from the frequency value, whereas the time information will be lost. Alternatively, to obtain the local features of the signal both in time and frequency domains, one shall perform STFT. This slices the signal into different segments using the window function $w(t)$ and each of these segments is subjected to FT as follows:

$$X(\tau, t) = X(t) \cdot w(t - \tau),$$

where t is the centre of the window function. The window function is placed such that the centre of the window coincides with the start of the signal and it traverses along the length of the signal. The following relations hold good:

$$X(\tau, \varepsilon) = \int X(\tau, t) e^{-j\varepsilon t} dt,$$

$$X(\tau, \varepsilon) = \int X(t) w(t - \tau) e^{-j\varepsilon t} dt,$$

where τ is the centre of the window in time and ε is the mean frequency of the window. $w(t - \tau) e^{-j\varepsilon t}$ is the STFT atom or the analysing function. To analyse the observed signal, it should match with this atom. For a specific condition where the analysing function is zero, it cannot extract any feature from the signal. Further, window should have compact support, which means that it should exist only over a finite time and vanishes outside the interval. If the window length is too long and equal to the length of the signal, then this process will converge to FFT. The inverse STFT is given by

$$X(t) = \frac{1}{2\pi} \int\!\!\int X(\tau, \varepsilon) e^{j\varepsilon t} d\varepsilon$$

$$= \frac{1}{2\pi} \int\!\!\int \langle X(t) \cdot w(t - \tau) e^{-j\varepsilon t} w(t - \tau) e^{-j\varepsilon t} d\varepsilon \, d\tau.$$

Spectrogram is the squared magnitude of the STFT. Spectrogram is the energy density in the time frequency plane. Energy decomposition of the signal is given by

$$\int |X(t)|^2 \, dt = \frac{1}{2\pi} \int |X(\omega)|^2 \, d\omega = \frac{1}{2\pi} \iint |X(\tau, \varepsilon)|^2 \, d\varepsilon \, d\tau.$$

The spectrogram $S(\tau, \varepsilon)$ is given by

$$S(\tau, \varepsilon) = |X(\tau, \varepsilon)|^2 = \left| X(t)\omega(t - \tau)e^{-j\varepsilon t} dt \right|^2.$$

4.28. Validation of the Developed SHM System

In order to validate the developed SHM system using WSN, the response of the scaled TLP is acquired using both wired system and wireless sensors. Results are compared in both time and frequency domains to estimate the error of disagreement, if any. Wired sensors are connected to the DAQ system through wires and the data are further processed using the central server, which is connected to the DAQ system. Wireless sensor nodes comprise of low-computing processor and sensor units. The processor shall acquire the data through MEMS-type sensors. Acquired data will be transmitted through Wi-Fi transmitter, which is connected to the SHM system. The central server is also connected to the same network, which receives the data transmitted by the wireless sensor nodes. Specifications of both the wired and wireless sensors are given in Table 4.7. The time-domain response of the TLP obtained by wired and wireless sensors are compared. Figure 4.28 shows the acceleration time history acquired by both wired and wireless sensors.

Table 4.7. Specifications of wireless accelerometer.

Description	Wired	Wireless
Accelerometer	393B04	MPU6050
Type	ICP	MEMS
Number of axis	1	3
Range	$\pm 5g$	$\pm 16g$ (opted $\pm 2g$)
Sensitivity	1 V/g	16,384 LSB/g
Noise performance	0.30 μg/$\sqrt{\text{Hz}}$	400 μg/$\sqrt{\text{Hz}}$

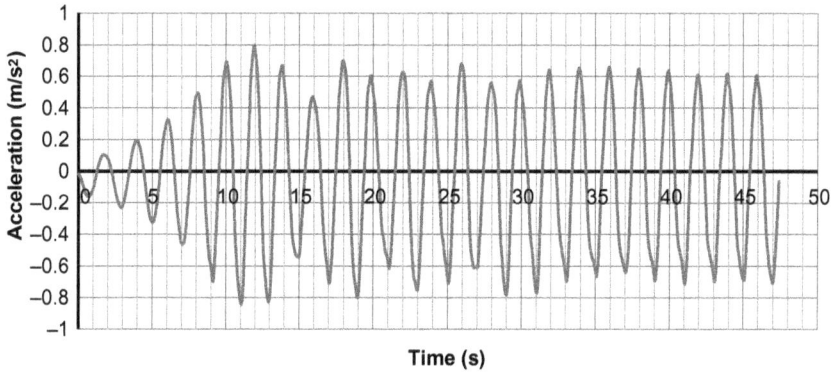

(a)

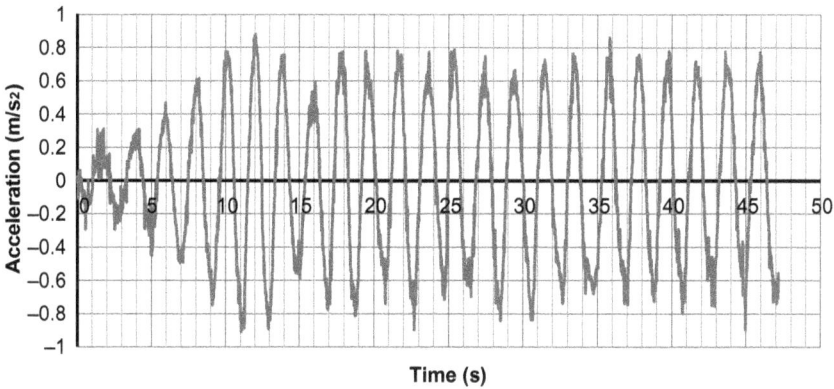

(b)

Figure 4.28. Acceleration response acquired by both sensors: (a) Wired acquisition; and (b) wireless acquisition.

Surge response of the deck is acquired by placing accelerometers on diametrically opposite locations on the deck. This is done to capture the influence of the wave direction on the DAQ of wireless sensors. The experiments are carried out with 8 and 10 cm wave heights in order to observe the variations in amplitude values. Wave heights are chosen to assess the sensitivity of the acquiring device and capability of detecting the changes in the variation of dynamic response. Figures 4.29 and 4.30 show the response amplitude operator (RAO) of surge displacement under regular waves for 8 and 10 cm wave heights, respectively. By comparing the time history

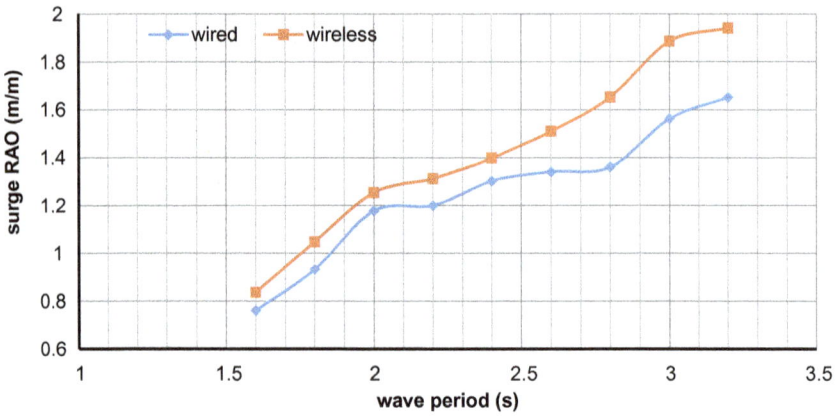

Figure 4.29. Surge RAO of TLP model (WH = 8 cm).

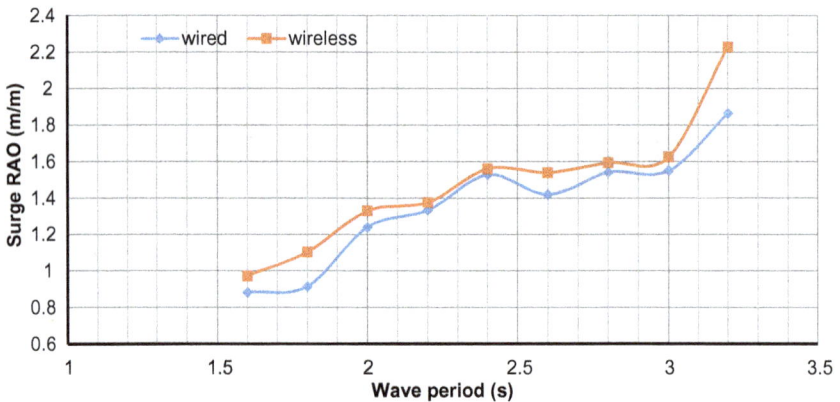

Figure 4.30. Surge RAO of TLP model (WH = 10 cm).

responses recorded by both types of sensors, it is seen that they show the same trend qualitatively for both sets of records. Data recorded using the newly proposed wireless sensor networking show marginal deviation for the length of data acquired. Table 4.8 shows the comparison of data. On comparison, it is seen that the percentage error is higher for higher periods and that the responses by the wired

Table 4.8. Comparison of surge response (RAO).

Wave period (s)	WH = 8 cm			WH = 10 cm			WH = 12 cm		
	Wired	Wireless	Difference (%)	Wired	Wireless	Difference (%)	Wired	Wireless	Difference (%)
1.6	0.76	0.84	10.00	0.89	0.98	10.18	0.89	0.97	9.71
1.8	0.94	1.05	12.28	0.92	1.01	9.58	0.99	1.13	13.67
2	1.12	1.26	6.53	1.25	1.33	7.15	1.18	1.28	8.59
2.2	1.20	1.32	9.43	1.34	1.38	3.26	1.31	1.44	9.36
2.4	1.31	1.40	7.33	1.54	1.57	2.17	1.34	1.53	14.59
2.6	1.34	1.51	12.62	1.42	1.55	8.47	1.42	1.55	9.73
2.8	1.36	1.66	21.41	1.54	1.59	3.44	1.34	1.73	29.77
3	1.57	1.89	20.70	1.55	1.63	4.87	1.53	1.98	29.82
3.2	1.56	1.94	17.57	1.87	2.22	19.50	1.58	1.96	23.46

and wireless systems have slight discrepancies. Marginal variation between the acquired values is mainly due to the sensitivity of the device as different sensors are used in wired and wireless systems. It is also due to the fact that different acquisition methods and processing techniques influence the variations. It is observed that the proposed wireless system is able to capture salient observations without missing out critical data points that are acquired by wired sensors. This shows a partial validation of the proposed WSN, which has many advantages: (i) easy processing of the acquired data; (ii) less time and space during installation and operation; and (iii) fast detection of the faulty sensor nodes.

PSD plots of surge response of the deck, obtained after post-processing the time history response are shown in Figure 4.31. It is seen from the figures that a marginal variation exists in terms of global features without time localisation. Peak frequencies of both wired and wireless SHM (wireless sensor 1) are acquired at 0.5335 and 0.553 Hz with a marginal difference of about 3.52%, keeping wired sensor data as the reference. In case of the second set of data acquired

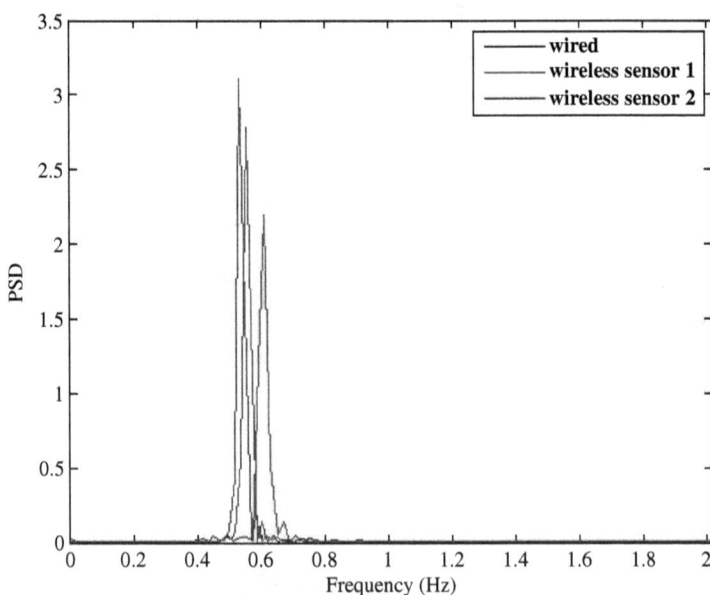

Figure 4.31. Frequency-domain response of deck in surge.

with WSN 2, the peak value is seen at 3.1 and 2.8 for wireless and wired, respectively, with an error of about 10%. Sensor 2, which is placed opposite to that of the wave approaching direction, showed a variation in the peak power value with respect to that of the wired sensor. A shift in the peak frequency is also observed. This shift is attributed to the time delay (lag) in the response of the platform with respect to that of the wave period. A set of sensors, placed at different locations on the deck, showed their sensitivity with respect to wave direction.

4.29. Reliability Assessment

Reliability assessment is expressed in terms of probability that a system will continue to perform its intended function under pre-stated operating conditions over a specified period of time. Reliability is estimated in statistical terms by monitoring the stability of various parameters operating under elevated stress conditions. Two approaches to derive reliability are either to evaluate probability of failure or probability of survival; the former is most commonly practiced in structural reliability. Probability of failure, as referred in the present context, is the probability that the response exceeds the threshold value, whereas the threshold value refers to the maximum amplitude acquired by exciting the scaled model under normal condition without any postulated failures. If the acquired responses under postulated failure cases show exceedance of the preset threshold value or an up-crossing over the threshold, it is considered to be an exceedance event.

4.29.1. *Assumptions*

Peak amplitude of the acquired response, in the respective degree of freedom, under the normal conditions without any postulated failures is assumed to be the threshold value. Response amplitude, acquired during the postulated failure cases, is considered to be in an alarming condition when it exceeds the threshold value. System failure is defined as user-induced postulated failures in the experimental model; progressive failure is not considered in this study.

4.30. Post-Processing Conditions

Statistics of the acquired acceleration time histories are quantified by their probability distribution. However, there exists some level of uncertainty in estimating the type of distribution and exact shape in comparison to that of the real structure. Since reliability estimates will be inaccurate if the distribution does not fit well with the set of acquired data, the best of fit is used to extrapolate the data beyond the range. In order to verify this fact, distribution analyses of the acquired data are performed and the probability plots are used for comparing the distribution. Goodness of fit test measures the compatibility of a random sample with a theoretical probability distribution function. In case more than one type of distribution fits the acquired data, an appropriate distribution with the lowest statistic value is considered as the best fitting model. On this basis, rank is assigned to the distributions to choose the most valid one. chi square goodness of fit statistics is used in the probability estimation, which is defined as follows:

$$\chi^2 = \sum_{i=1}^{k} \frac{(O_i - E_i)^2}{E_i},$$

where O_i is the observed frequency for the ith bin and E_i is the corresponding expected frequency and is given by

$$E_i = F(x_2) - F(x_1),$$

where F is the cumulative distribution function (CDF) of the probability distribution being tested, and $\{x_1, x_2, \ldots, x_i\}$ are the limits for the ith bin. The probability of exceedance is calculated for each postulated failure case and listed in Table 4.9.

4.31. Frequency-Domain Response

PSD value of surge response for various cases of postulated failure is presented in Figure 4.32. The platform is subjected to regular waves of wave height of 8 cm and period of 2 s. By comparing the plots of postulated failure cases with that of the normal cases, it is seen that the frequency at which peak surge response occurs has

Table 4.9. Probability distribution and value of exceedance.

Postulated failure case	DOF	Distribution	Parameters	Probability of exceedance (%)
Case 1	Surge	Cauchy	$\mu = -0.0109$ $\sigma = 0.0642$	4.22
	Pitch	Cauchy	$\mu = 0.0234$ $\sigma = 0.3702$	7.41
	Heave	Dagum (4p)	$\alpha = 15.382$ $\beta = 0.4326$ $\gamma = -0.3340$ $k = 0.303$	40.07
Case 2	Surge	Cauchy	$\mu = 0.0181$ $\sigma = 0.0801$	5.58
	Pitch	Cauchy	$\mu = 0.3445$ $\sigma = 0.5010$	12.22
	Heave	Gen extreme value	$\mu = -0.0147$ $\sigma = 0.0266$ $k = -0/1546$	2.68
Case 3	Surge	Cauchy	$\mu = -0.0133$ $\sigma = 0.0772$	5.03
	Pitch	Cauchy	$\mu = 0.0106$ $\sigma = 0.1131$	2.28
	Heave	Log logistic	$\alpha = 14.006$ $\beta = 0.4893$ $\gamma = -0.4861$	18.13
Case 4	Surge	Cauchy	$\mu = -0.0119$ $\sigma = 0.1027$	6.63
	Pitch	Cauchy	$\mu = -0.0466$ $\sigma = 0.2173$	4.21
	Heave	Cauchy	$\mu = 0.0159$ $\sigma = 0.0341$	21.43
Case 5	Surge	Cauchy	$\mu = 0.0204$ $\sigma = 0.4765$	25.91
	Pitch	Cauchy	$\mu = -0.0173$ $\sigma = 0.8381$	15.34
	Heave	Dagum (4p)	$\alpha = 13.815$ $\beta = 0.4665$ $\gamma = -0.3864$ $k = 0.3380$	30.3

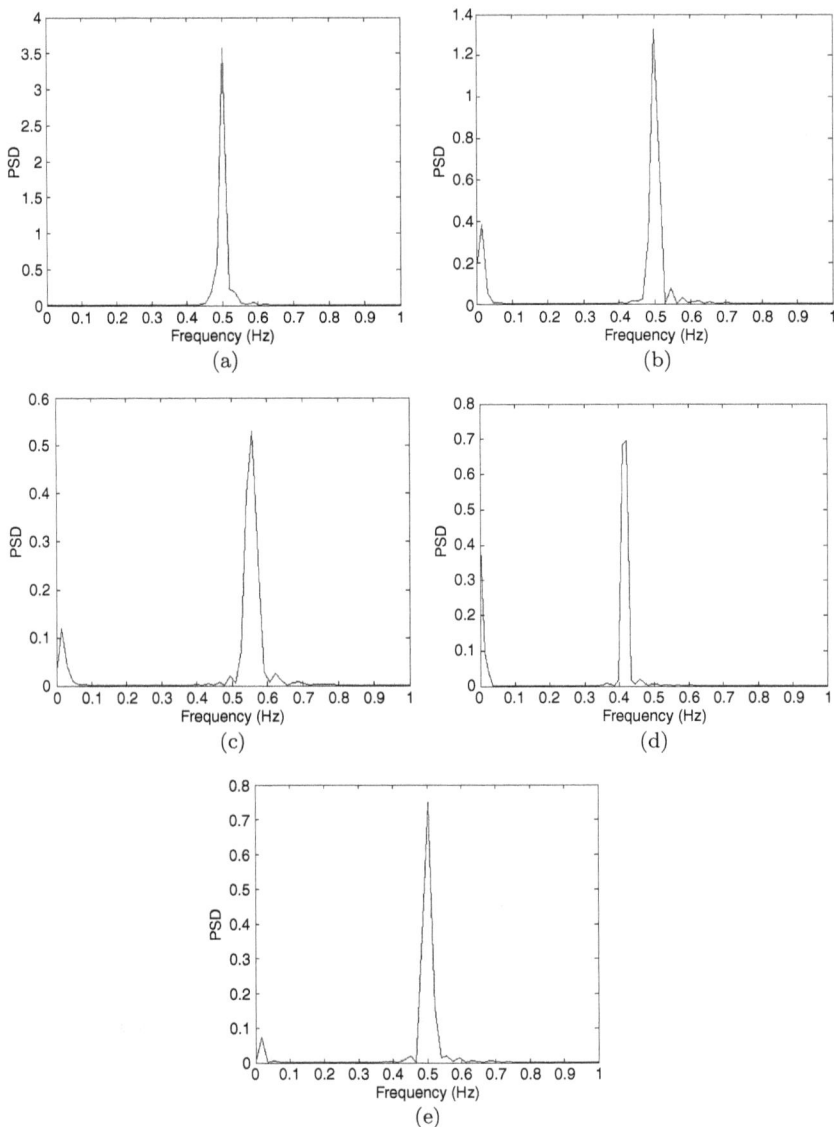

Figure 4.32. PSD value of surge response: (a) normal case; (b) postulated failure (case 1); (c) postulated failure (case 2); (d) postulated failure (case 3); and (e) postulated failure (case 4).

insignificant variation. However, smaller peaks seen near to 0.5 Hz in the postulated failure cases are mainly due to the second-order vibrations caused by tether failure of respective legs. As TLPs are designed to remain highly flexible (compliant) in the horizontal plane, second-order harmonics are causing local response in frequency bands closer to that of the peak frequency. A significant peak appearing closer to the origin indicates a strong coupling effect of surge with that of heave. In all cases, the peak frequency remains more or less same with a negligible variation. A small peak seen at the origin, which is significantly high for postulated failure cases, is a clear manifestation of postulated failure, which is captured by the sensor network.

Pitch response for damaged and normal case remains the same with a marginal variation in the amplitude. Figure 4.33 shows the pitch response for various cases. It is seen from the figures that the chosen sensors are capable of registering critical response characteristics in terms of frequency contents and amplitude without any loss of packet of signals. Figure 4.34 shows the heave response for the normal and postulated failure cases. Heave response of the normal case has various peaks: first peak, occurring at zero frequency is significantly low; second peak occurs closer to the wave excitation frequency; and successive smaller peaks occur closer to the heave frequency. But the damaged cases show a significantly high magnitude of heave responses at higher frequencies. Shifts of the first few peaks, in terms of both amplitude and frequency of occurrence, are due to the eccentric load over column 2. Postulated failure cases showed clear manifestation of failure, which is satisfactorily captured by the proposed WSN. With the PSD plots, only global features of the postulated damages are extracted without any localisation in the time domain. To get the local features of frequency across the time value, STFT results are plotted for a chosen hamming window.

Different scales on $X-$ and Y-axis are chosen for the PSD plots in order to capture the entire amplitude value of the PSD. With the same scale on both the abscissa and ordinates, some magnitudes

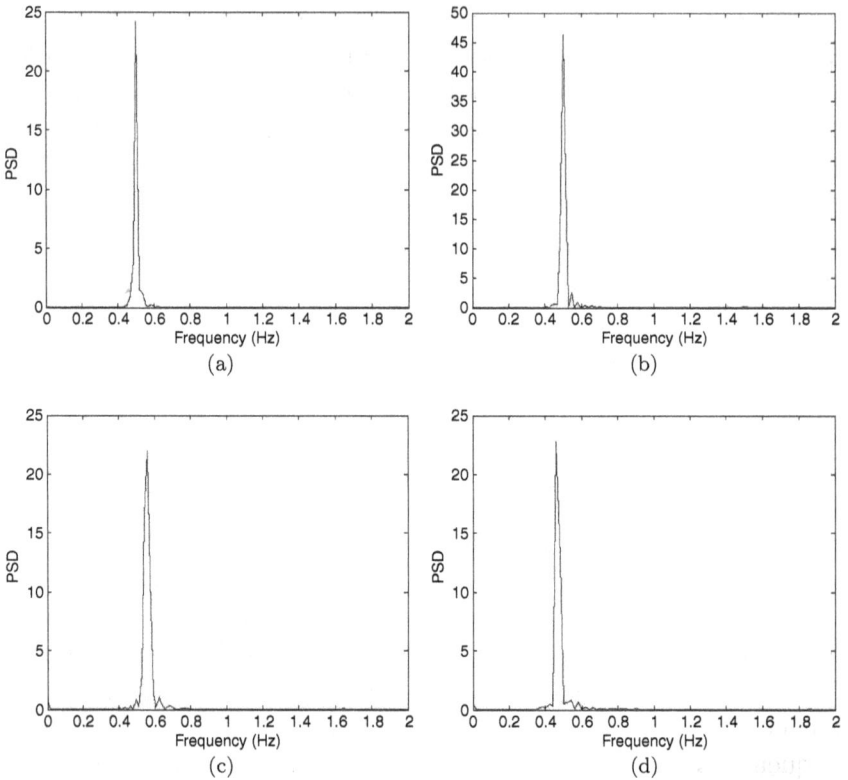

Figure 4.33. PSD value of pitch response: (a) normal (case); (b) postulated failure (case 1); (c) postulated failure (case 2); and (d) postulated failure (case 3).

will not be significant in comparison. Figure 4.35 shows 3D graph for STFT of surge response under normal and postulated damaged cases. It is seen from the plots that a wide range of frequency variations occurring throughout the time history are successfully acquired. These data are useful in diagnosing the condition of the platform for further assessment.

Further, Figure 4.36 shows heave responses under normal, postulated failure cases 2, 4 and 5 under eccentric load over position 4, removal of tether at position 4 and under extreme wave

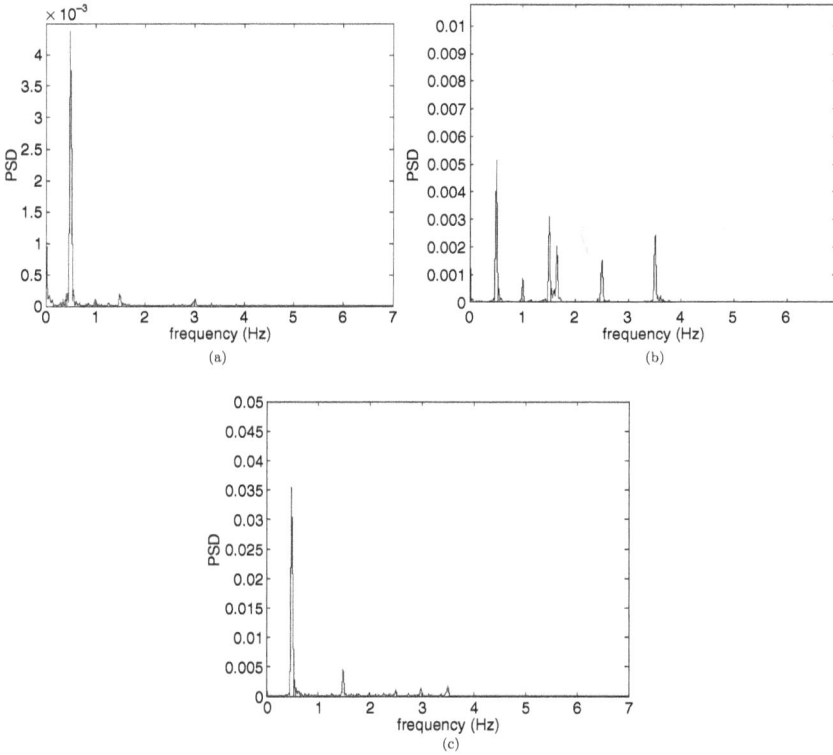

Figure 4.34. PSD of heave response: (a) normal (case); (b) postulated failure (case 1); and (c) postulated failure (case 4).

height, respectively. Large set of frequencies and the corresponding variations along the timescale are successfully captured by the WSN.

The platform when excited by the unidirectional wave should not have a roll response. When the eccentric load is added on position 1, the offshore TLP model under postulated failure case 1 is observed to have a significant roll response as shown in Figure 4.37. A wide spectrum of frequency distribution under the postulated case shows the necessity of DAQ for a wide bandwidth to capsulate the efficient alert monitoring of the platform. A large set of frequencies and the corresponding variations along the timescale are successfully captured by the WSN.

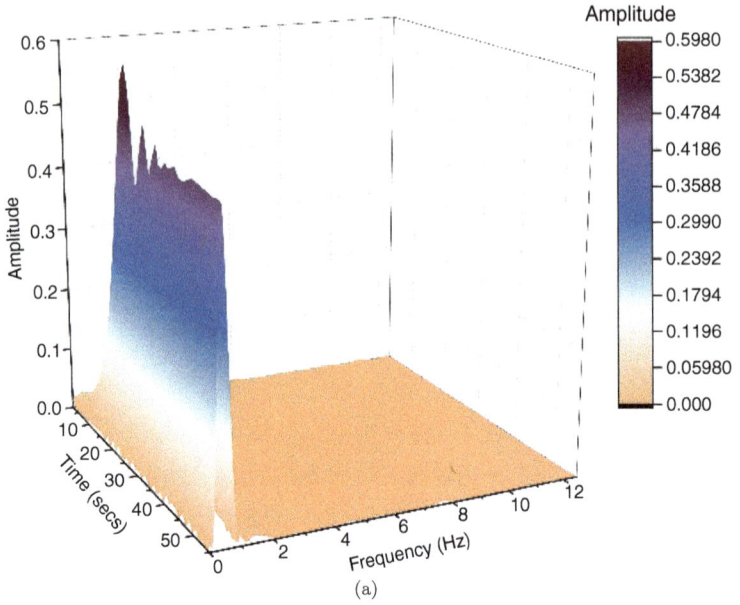

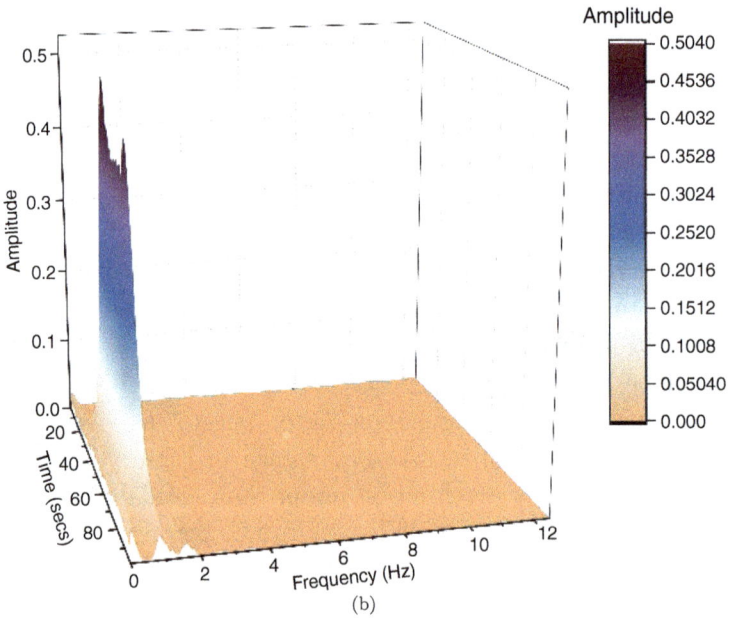

Figure 4.35. Short-time Fourier transform of the surge response: (a) normal case; and (b) postulated failure (case 2).

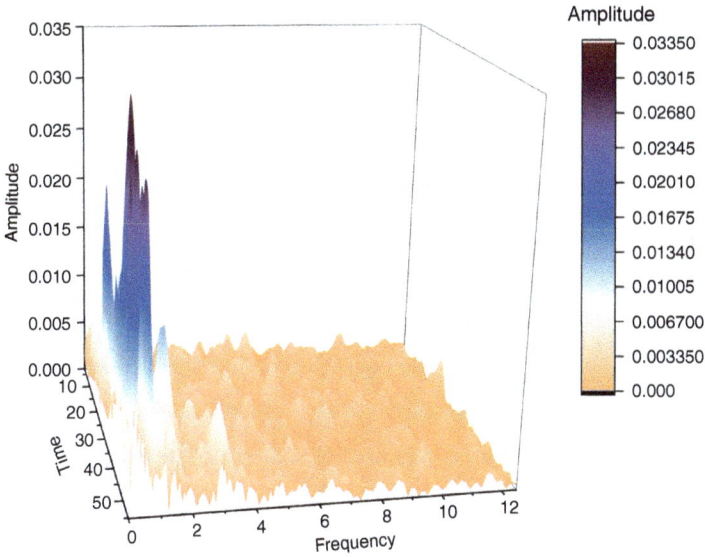

(a)

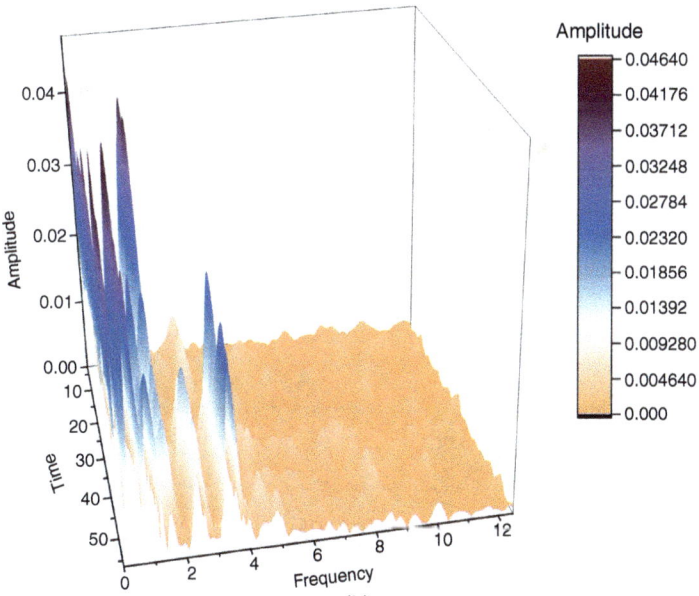

(b)

Figure 4.36. Short-time Fourier transform of the heave response: (a) normal case; and (b) postulated failure (case 2);

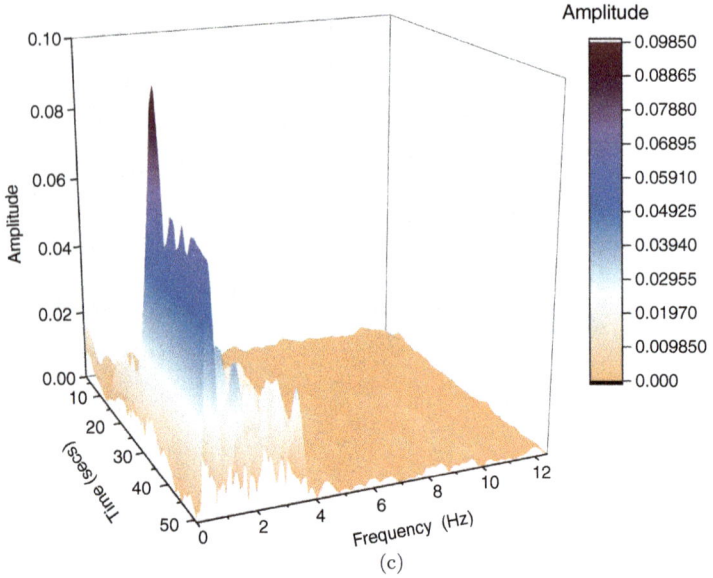

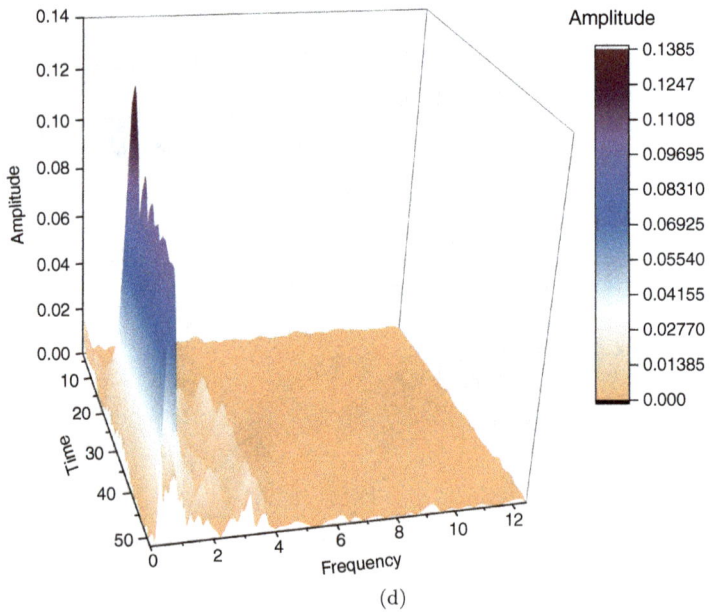

Figure 4.36. (*Continued*) (c) postulated failure (case 4); and (d) postulated failure (case 5).

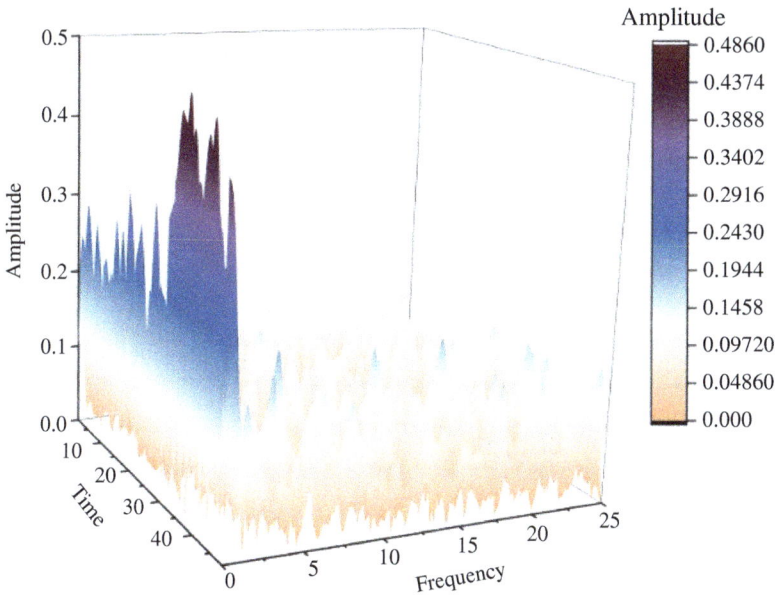

Figure 4.37. Short-time Fourier transform of the roll response for postulated failure (case 1).

4.32. User Interface and Alert Triggering System

Data acquired by sensor nodes are transmitted to the base station and stored in a MySQL database. Acquired data are processed and enabled for viewing as a report at the user interface. Acceptable limit of the acceleration value is fixed as the threshold value; in this study, the response of the platform under the normal case is considered as the base value for comparison, which can be changed in the user interface, if desired. On exceedance of the preset threshold values, an alert message will be displayed at the user interface. In a real case, the layout of sensor nodes will be integrated with the database. After processing the data, the central server will send SMS to the registered mobile number on exceedance of the threshold value. With the SMS-API, SMS is triggered from the website for two-factor authentication. Figure 4.38 shows the user interface with the alert message displayed on top of the home screen.

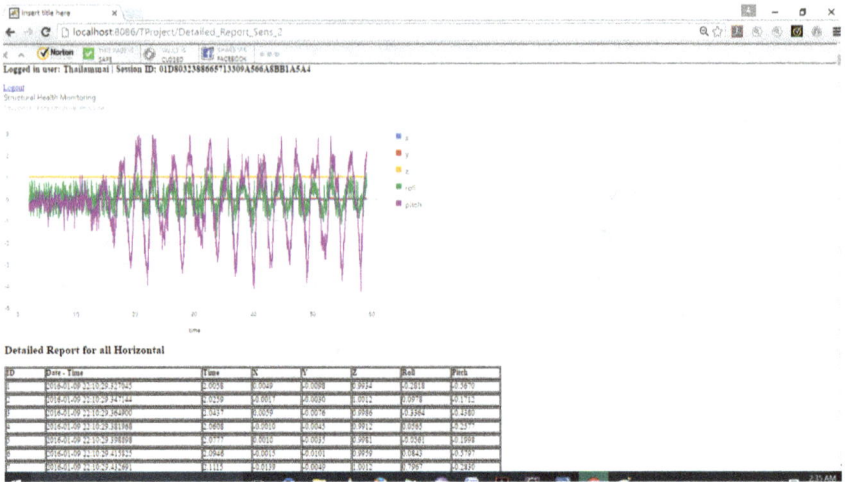

Figure 4.38. Screen shot of the user interface.

4.33.   Health Monitoring of BLSRP in Lab Scale

Experimental investigations carried out on a scaled model of BLSRP to assess the indigenously designed WSN system and SHM scheme are presented. The primary idea is to investigate the successful application of the proposed architecture of WSN, which is then subsequently applied to analyse postulated failure cases of the offshore platform. An extension of AMS is also proposed to make use of the proposed WSN in terms of successful SHM. Sensor node with ARMv7 processor and MPU6050 sensor unit with IEEE 802.11 protocol are deployed in the SHM system. Health monitoring system used in this study is of nomenclature SHM 2.

BLSRP is a new innovative structure in which the transfer of rotational degrees of motion from buoyant legs to the deck is restrained by hinged joints. Figure 4.39 shows the experimental setup. The deck of the platform rests on buoyant legs, which in turn are anchored to the sea bed using taut-moored tethers. Deck and the BLS are connected by the hinged joints, which allow the transfer of translation motion from the buoyant legs to the deck but restrain the transfer of rotations. While buoyant legs act as storage units, BLSRP is mainly used to process LNG before it is exported to shore. A scaled

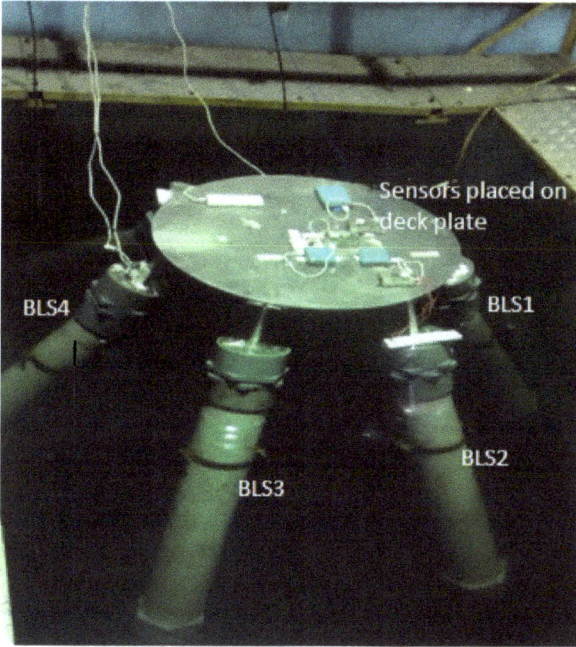

Figure 4.39. Experimental setup of BLSRP with sensor nodes.

(1:150) model used for this study is fabricated as follows: (i) deck and hinged joints are made of stainless steel; (ii) buoyant legs are made of PVC material; and (iii) tethers of mild steel are used. Buoyant legs are ballasted using sand to attain the required draft level. The base plate is fixed at the bottom of the sea bed to which the tethers are connected using rollers; a top-tension system is used to impose pre-tension in tethers. One end of the tether is connected to each of the buoyant leg while the other end is connected to the seabed. Structural details of the platform are given in Table 4.10.

4.34. Implementation of SHM System in BLSRP: Lab Scale

Sensor boards are placed both on the deck and buoyant legs as shown in Figure 4.39. Both wired and wireless accelerometers are placed to compare the acquired responses under unidirectional regular

Table 4.10. Structural properties of BLSRP.

Description	Prototype	Model 1:150
Mass of the structure	400000 ton	118.51 kg
Diameter of the deck	22.5 m	150 mm
Length of each leg	200 m	1333.33 mm
Diameter of each leg	22.5 m	150 mm
Draft	163.57 m	1117.73 mm
Ballast	333950 tons	98.94 kg
Length of the tether	470.84 m	1333.33 mm
Diameter of the tether	0.05 m	0.33 mm
Water depth	600 m	4000 mm

Figure 4.40. Underwater accelerometer fixed to buoyant leg.

waves. Underwater accelerometer is placed on the leg below the still water level (Figure 4.40). Detailed specifications of the underwater accelerometer are given in Table 4.9. Load cells are connected to tethers to acquire dynamic tether tension variations during the course of experiments. The model is excited by regular waves with wave height of 8 to 12 cm and period ranging from 1.6 to 3.2 s. DAQ of both the types of sensors is carried out simultaneously. Wireless sensor nodes of SHM system are placed at the Cg of the deck and in all buoyant legs. Based on the postulated failure cases, the position of sensors are changed so that the dominant effect of damage can be captured. For some cases, different trials are carried out by placing the sensors in all buoyant legs to choose the most appropriate location of sensor deployment. Each set of experiments is repeated for a minimum of three trials and the average of all trials is considered for data processing.

While surge and heave motions of the deck are measured using both wired and wireless accelerometer modules, motions of that of the buoyant legs are measured using underwater accelerometer module 393B04. This is a specially designed module consisting of insulated components with water-proof capabilities and can be fixed underwater. A rigid clamp is fixed around each of the buoyant legs to hold the underwater accelerometers. As it is uniaxial, the direction of measurement is changed according to the degree of freedom to be measured; experiments are repeated for same set of wave loads to measure different motions of the buoyant legs under unidirectional waves of different wave heights and periods. Specifications of the underwater accelerometer module used in the study are given in Table 4.11. Based on the free-vibration tests carried out on the scaled model of BLSRP, natural periods in surge, heave, and pitch are found to be 9.21 s, 0.28 s and 0.35 s, respectively; damping ratios are 8.12%, 3.5%, and 6.63%.

Table 4.12 gives the detailed specifications of accelerometer deployed during the experimental investigations.

Table 4.11. Specifications of underwater accelerometer.

Description	Wired
Underwater accelerometer	W393B04
Type	Seismic miniature; flexural ceramic; piezoelectric integrated circuit
Number of axis	One
Range	$\pm 5g$
Sensitivity	1 V/g
Noise performance	0.30 $\mu g/\sqrt{Hz}$

Table 4.12. Accelerometer specifications.

Parameter	Conditions	Minimum	Type	Maximum	Units
Accelerometer Sensitivity					
Full-scale range	AFS_SEL = 0		±2		g
	AFS_SEL = 1		±4		g
	AFS_SEL = 2		±8		g
	AFS_SEL = 3		±16		g
ADC word length	Output in two's complement format		16		bits
Sensitivity scale factor	AFS_SEL=0		16,384		LSB/g
	AFS_SEL = 1		8192		LSB/g
	AFS_SEL = 2		4096		LSB/g
	AFS_SEL = 3		2048		LSB/g
Initial calibration tolerance			±3		%
Sensitivity change vs. temperature	AFS_SEL = 0, −40°C to +85°C		±0.02		%/°C
Nonlinearity	Best fit straight line		0.5		%
Cross-axis sensitivity			±2		%
Zero-G Output					
Initial calibration tolerance	X- and Y-axes		±50		mg
	Z-axis		±80		mg
Zero-G level change vs. temperature	X- and Y-axes, 0°C to +70°C		±35		
	Z-axis, 0° C to +70°C		±60		mg
Self-Test Response					
Relative	Change from factory trim	−14		14	%
Noise Performance					
PSD	@10 Hz, AFS_SEL = 0 & ODR = 1 kHz		400		$\mu g/\sqrt{Hz}$
Low Pass Filter Response					
Programmable range		5		260	Hz
Output Data Rate					
Programmable range		4		1,000	Hz
Intelligence function increment			32		mg/LSB

4.35.   Validation of SHM System in BLSRP Model

To validate the proposed wireless SHM system, the response of the model is obtained using both wireless and wired sensor systems. Response of the model is acquired under the action of unidirectional regular waves of 10 cm wave height; wave period is varied from 1.6 to 3.4 s, which can be generated by the wave flume with a correct pattern of sine wave. The time-domain response of the deck under regular wave (10 cm, 1.8 s) is shown in Figure 4.41. Results are compared in both time and frequency domains to diagnose the discrepancy. Wired sensors are connected to the DAQ system and data are post-processed using a central server. Wireless sensor nodes comprise of low-computing processor and sensor units. The processor acquires the data through MEMS-type sensors, which is further transmitted through Wi-Fi transmitter. The central server is also connected to the same network, which receives the data transmitted by the wireless sensor nodes as well. Figure 4.42 illustrates RAO values of deck in surge, heave and pitch degrees of freedom under the regular waves of 10 cm wave height; Figure 4.43 shows that of the buoyant leg. Figures 4.44 and 4.45 show comparison of responses of the deck in surge and pitch degrees-of-freedom, respectively. It is seen from the figures that data acquired by both set of sensors in both translational and rotational degrees of freedom are qualitatively similar; this is true for both deck and buoyant leg responses, which confirm the effective integration of underwater sensor in the sensor network. Comparison of the responses is also listed in Tables 4.13 and 4.14, respectively, which shows a good agreement between the data acquired by wired and wireless sensors. It ensures that the wireless SHM system is capable of capturing the salient observations without missing out the critical data points that are acquired by wired sensors. Marginal variations, as seen in the values, are attributed to the range of sensitivity as different types of sensors are deployed in wired and wireless systems. Different acquisition methods and processing techniques also influence these variations.

Responses in rotational degrees of freedom are acquired using the wired inclinometer. In case of the wireless SHM, MPU6050 is

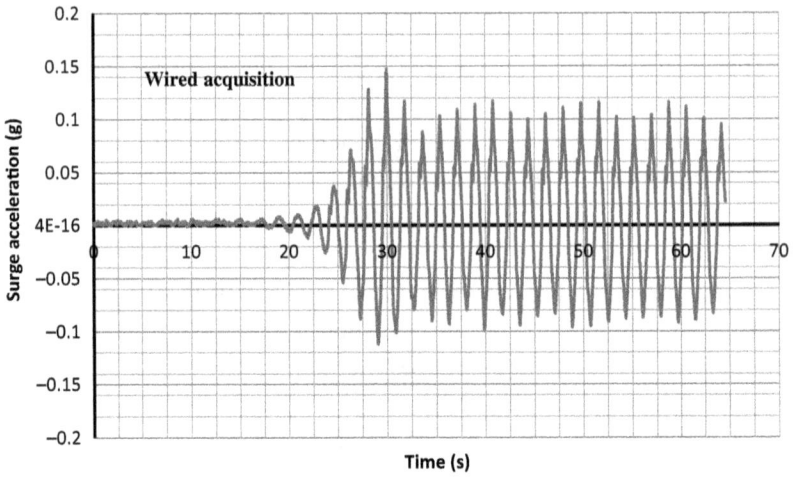

(a)

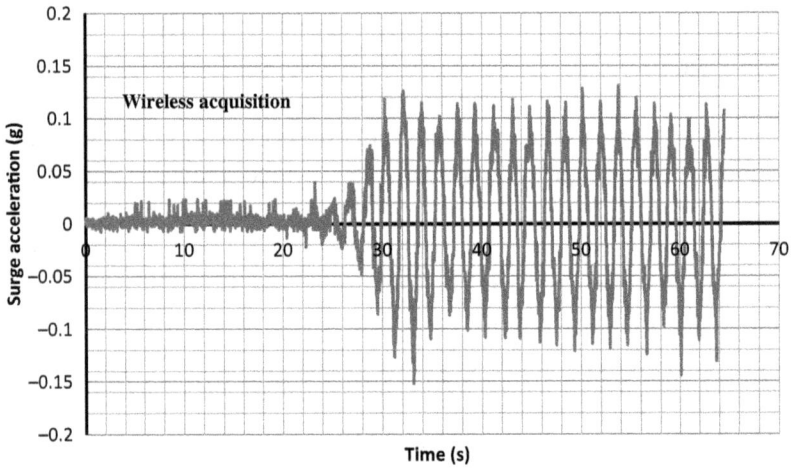

(b)

Figure 4.41. Deck response of BLSRP: (a) wired sensors; and (b) wireless sensors.

deployed, which is a combination of gyroscope and accelerometer. PSD plots of surge and pitch response of the deck obtained after post-processing time history responses are shown in Figures 4.42 and 4.43, respectively. It is seen from the figures that the peak frequency of surge response occurs at 0.624 and 0.56 Hz in case of both wired

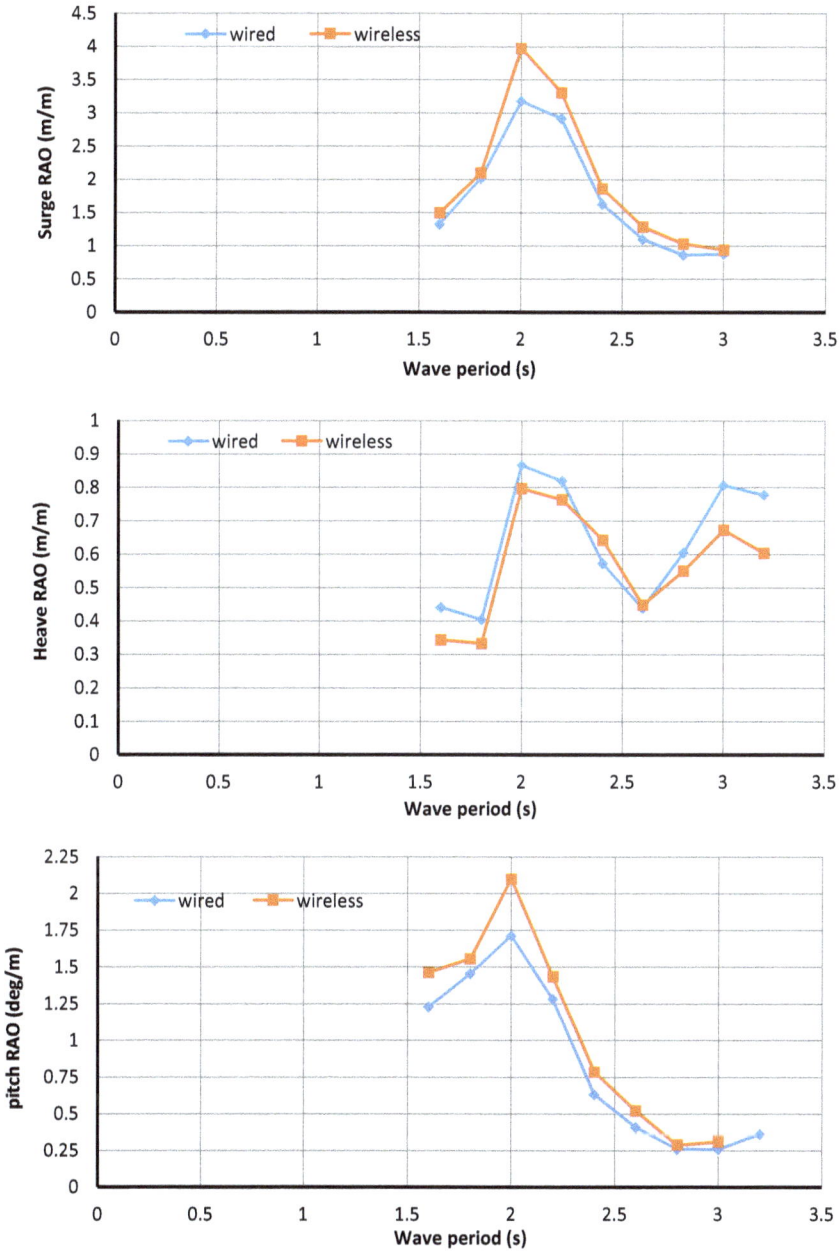

Figure 4.42. RAO of deck response: BLSRP.

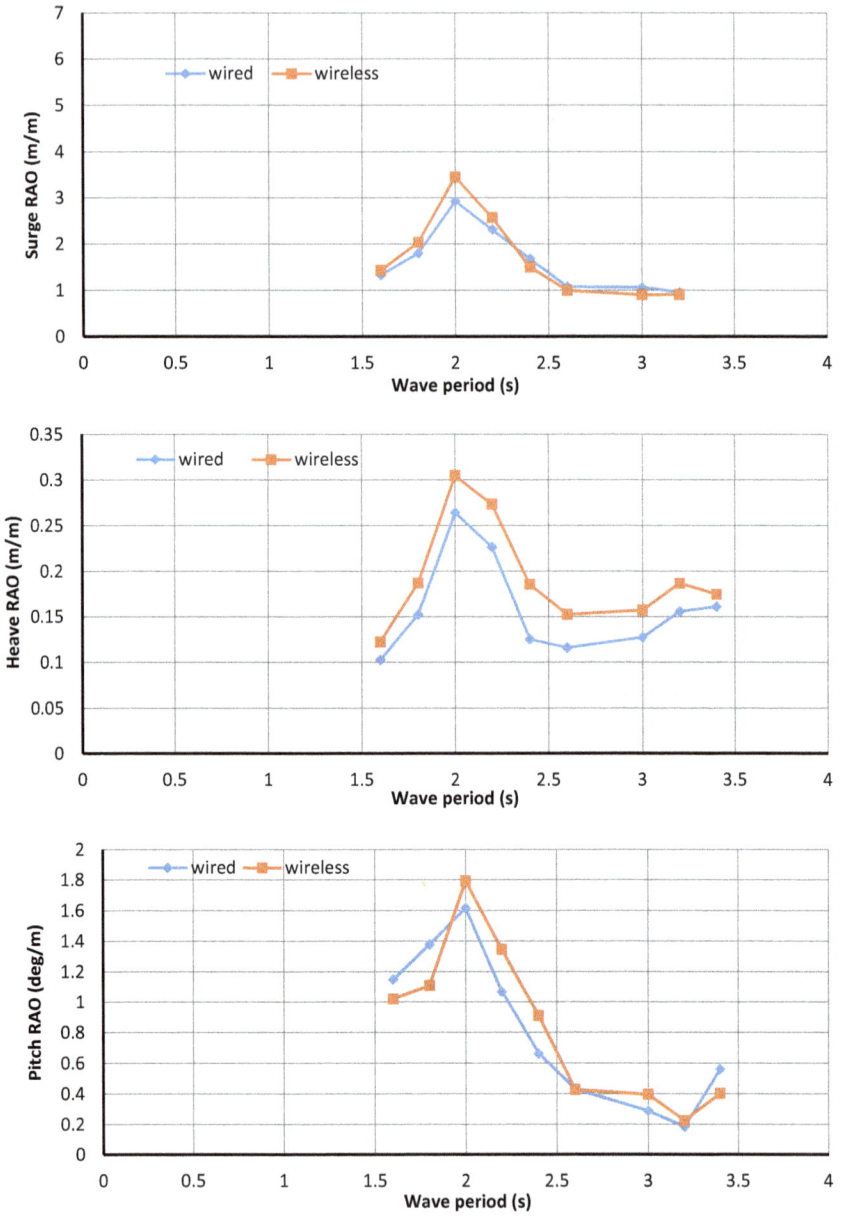

Figure 4.43. RAO of buoyant leg of BLSRP.

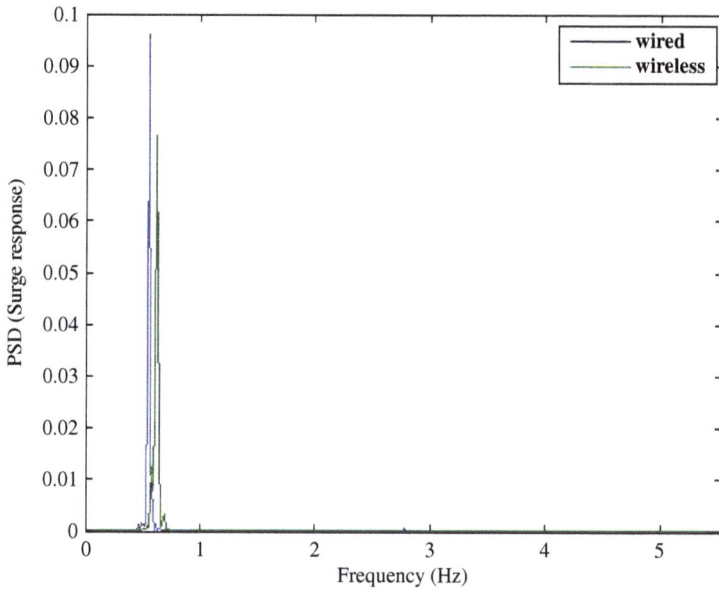

Figure 4.44. Comparison of surge response of BLSRP deck.

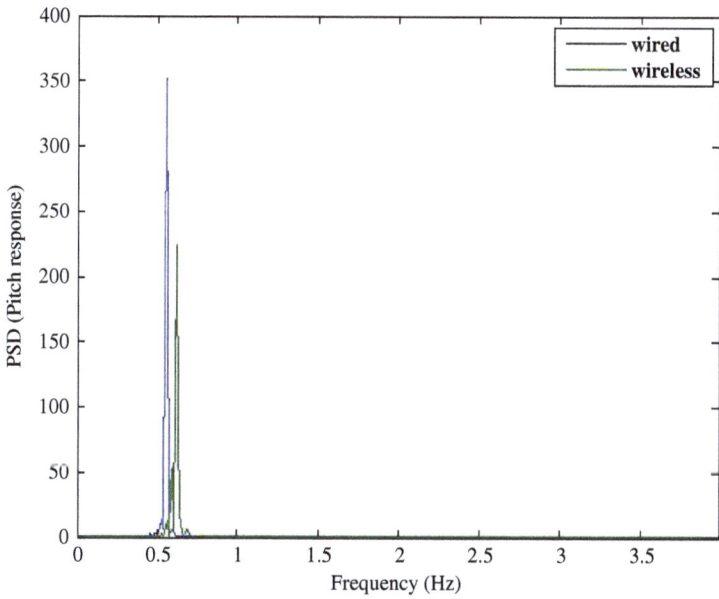

Figure 4.45. Comparison of pitch response of deck.

Table 4.13. Comparison of deck response of BLSRP.

Wave period (s)	Surge			Heave			Pitch		
	Wired	Wireless	Difference (%)	Wired	Wireless	Difference (%)	Wired	Wireless	Difference (%)
1.6	1.32	1.49	12.95	0.44	0.34	−22.08	1.22	1.46	19.25
1.8	2.01	2.09	4.06	0.40	0.33	−17.68	1.45	1.56	6.77
2.0	3.17	3.96	24.89	0.87	0.79	−8.14	1.71	2.1	22.64
2.2	2.91	3.30	13.22	0.82	0.77	−6.81	1.28	1.43	11.81
2.4	1.62	1.85	14.51	0.57	0.64	11.93	0.63	0.78	24.40
2.6	1.09	1.28	17.23	0.44	0.45	2.33	0.41	0.52	26.86
2.8	0.86	1.02	19.09	0.60	0.55	−9.09	0.26	0.29	9.912
3.0	0.87	0.93	6.923	0.81	0.67	−16.70	0.26	0.31	19.65

Table 4.14. Comparison of buoyant leg response of BLSRP.

Wave period (s)	Surge			Heave			Pitch		
	Wired	Wireless	Difference (%)	Wired	Wireless	Difference (%)	Wired	Wireless	Difference (%)
1.6	1.35	1.44	7.49	0.10	0.23	18.89	1.15	1.03	11.08
1.8	1.81	2.04	12.63	0.15	0.19	22.58	1.38	1.11	19.37
2.0	2.93	3.46	17.80	0.27	0.44	14.94	1.62	1.80	11.01
2.2	2.32	2.58	10.90	0.23	0.27	20.54	1.07	1.35	25.81
2.4	1.69	1.51	10.59	0.13	0.19	47.91	0.66	0.91	37.39
2.6	1.09	1.01	8.00	0.12	0.15	31.03	0.43	0.43	1.15
2.8	1.08	0.91	15.21	0.13	0.16	23.04	0.29	0.40	36.40
3.0	0.97	0.92	5.00	0.16	0.19	20.00	0.19	0.23	22.30

and wireless sensors, respectively. This shows a difference of about 11.4% in the frequency while that of the peak magnitude differs by about 14.44% in surge degree of freedom. Pitch response shows a difference of about 11.5% in the acquired frequency.

4.36.   Response of Buoyant Legs of BLSRP

Comparison of heave responses of the deck and the buoyant legs that are acquired using underwater accelerometer is shown in Figure 4.46. Two sensors are deployed to measure the response on buoyant leg (BLS 4); one is deployed just above the water level and one below. Both the sensors are capable of acquiring responses in a comparable accuracy. It is seen from the plots that peaks occurring at very low frequency and 0.5 Hz show coupling of heave with that of surge. It is also seen that the deck response is minimum compared to that of the buoyant legs, which is attributed to the presence of ball joints. This is evident by comparing the magnitude of heave responses at 0.5 Hz; the deck response is significantly lesser than that of the buoyant leg.

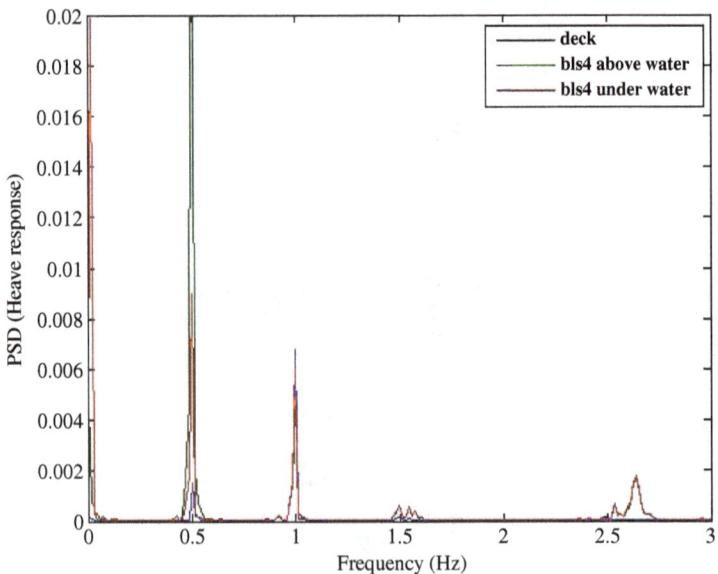

Figure 4.46. Comparison of heave response of BLSRP.

This ensures that the minimum effect of waveload is transferred from the buoyant leg to the deck as the desk is isolated from the buoyant leg by hinged joints. It is important to note that both sets of sensors deployed in the study are effective in their acquisition as they are capable of acquiring critical data without any loss.

4.37. Postulated Failure Cases

One of the main objectives of design and development of the SHM system is to examine its effectiveness in health monitoring under the postulated failure cases. Experimental investigations of BLSRP are carried out for a few postulated failure cases, based on which an appropriate AMS is proposed. The acquired response of both the deck and buoyant legs under normal sea state for regular waves are set as the threshold values of respective members. Platform is then subsequently examined under different postulated failure cases to assess the adaptability and efficiency of the proposed SHM network. Postulated failure case 1 refers to applying eccentric load on one of the buoyant legs, BLS 1, and is schematically shown in Figure 4.47. Appropriate steel blocks are added to the buoyant leg, BLS 1, by about 17% of its weight to lower its draft level. Postulated failure case 2 refers to the tether pull out of the buoyant leg, BLS 1, while that of case 3 refers to the response of the platform under extreme waves. Postulated failure cases are created to demonstrate the effectiveness of AMS. Monitoring system in this discussion is concerned only with the acceleration response of the platform. For exceedance of the acquired acceleration in comparison with that of undamaged case, it is designed to raise an alarm as part of the proposed SHM.

4.38. Reliability Formulation

Reliability is estimated in statistical terms by monitoring the stability of various parameters operating under elevated stress conditions. Two approaches to derive reliability are either to evaluate the probability of failure or probability of survival; the former is most commonly practiced in structural reliability. Probability of failure, as

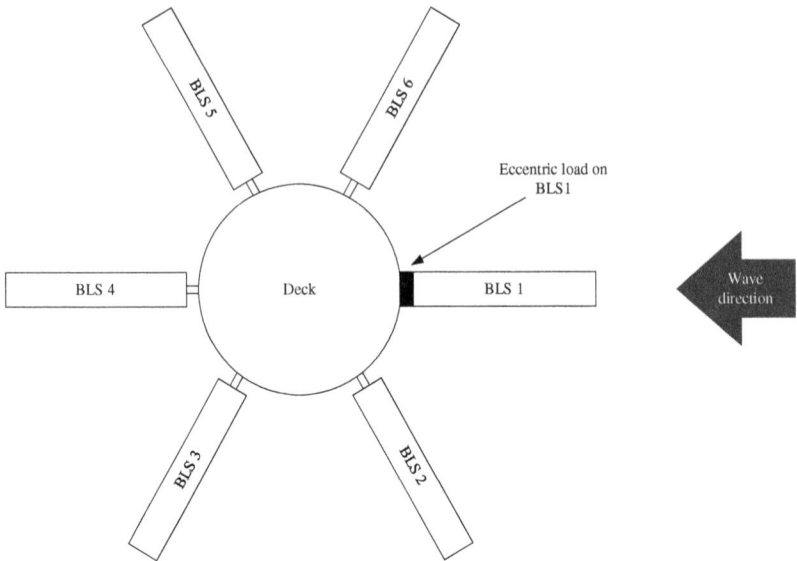

Figure 4.47. Postulated failure case 1.

referred in the present context, is the probability where the response exceeds the threshold value whereas the threshold value refers to the maximum amplitude acquired by exciting the scaled model under normal conditions without any postulated failures. If the acquired responses under postulated failure cases show exceedance of the preset threshold value or an up-crossing over the threshold, it is considered to be an exceedance event.

Probability distributions for the response of the platform in surge, heave and pitch are given in Table 4.15. It is seen from the table that there exist cases where the response, even under the postulated failure cases, do not exceed the threshold value; on the other hand, exceedance is very insignificant. A sample procedure is shown for one case; similar procedure is followed for all other cases. Let us consider the surge displacement of buoyant leg (BLS 1) under postulated failure case 1. As seen from the table, this has a triangular probability distribution, as shown in Figure 4.48, while the corresponding cumulative distribution is shown in Figure 4.49. Parameters for the triangular distributions, as seen from the table,

Table 4.15. Probability distribution and value of exceedance.

Postulated failure	Structural member and DOF	Distribution	Parameters	Probability of exceedance (%)
Case 1	BLS 1 Surge	Triangular	$m = 0.029$	1.27
			$a = -0.08449$	
			$b = 0.1204$	
	BLS 1 Heave	Gen gamma (4p)	$\alpha = 0.1426$	28.67
			$\beta = 0.1468$	
			$\gamma = -0.0710$	
	BLS 1 Pitch	Triangular	$m = 1.888$	1.44
			$a = -5.2518$	
			$b = 7.146$	
	Deck Surge	Weibull	$\alpha = 2.1395$	22.47
			$\beta = 0.1246$	
			$\gamma = -0.1103$	
	Deck Heave	Cauchy	$\mu = 3.9E{-}4$	30.54
			$\sigma = 0.0251$	
	Deck Pitch	Gen gamma (4p)	$\alpha = 0.0797$	24.8
			$\beta = 10.92$	
			$\gamma = -5.8731$	
Case 2	BLS 4 Surge	Gumbel max	$\mu = -0.0013$	0.32
			$\sigma = 0.0186$	
	BLS 4 Heave	Cauchy	$\mu = -0.0043$	4.64
			$\sigma = 0.0077$	
	BLS 4 Pitch	Cauchy	$\mu = -0.5957$	6.39
			$\sigma = 1.3791$	
	Deck Surge	Log logistic	$\alpha = 6.8045$	16.52
			$\beta = 0.1451$	
			$\gamma = -0.1391$	
	Deck Heave	Laplace	$\mu = -4.08{*}E{-}4$	21.69
			$\lambda = 45.356$	
	Deck Pitch	Triangular	$m = 0.1646$	5.86
			$a = -6.7628$	
			$b = 4.314$	
Case 3	BLS 1 Surge	Gumbel max	$\mu = -0.0056$	14.3
			$\sigma = 0.05924$	
	BLS 1 Heave	Cauchy	$\mu = 0.0124$	17.98
			$\sigma = 0.02458$	
	BLS 1 Pitch	Laplace	$\mu = 0.1848$	4.93
			$\lambda = 0.3866$	
	Deck Surge	Cauchy	$\mu = -8.03\,E{-}4$	10.73
			$\sigma = 0.01374$	
	Deck Heave	Cauchy	$\mu = -0.00136$	8.28
			$\sigma = 0.0044$	
	Deck Pitch	Cauchy	$\mu = -0.0372$	14.97
			$\sigma = 1.3774$	

Probability Density Function

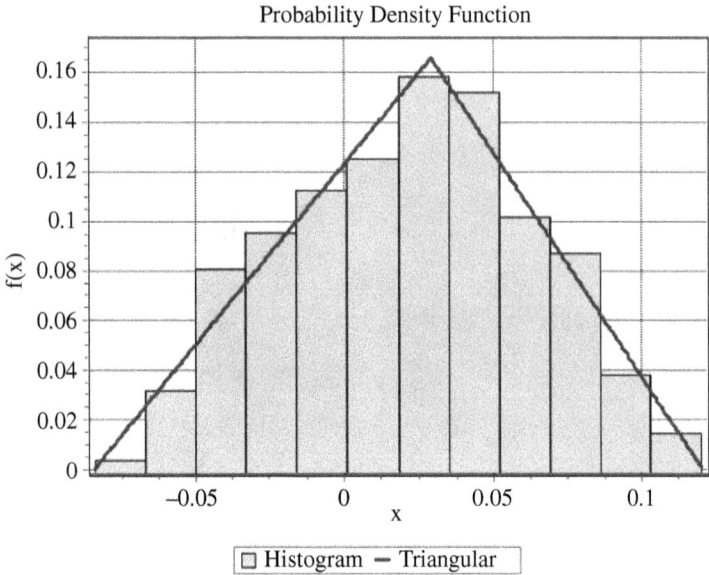

Figure 4.48. Probability distribution for BLS 1 for surge response (case 1).

Cumulative Distribution Function

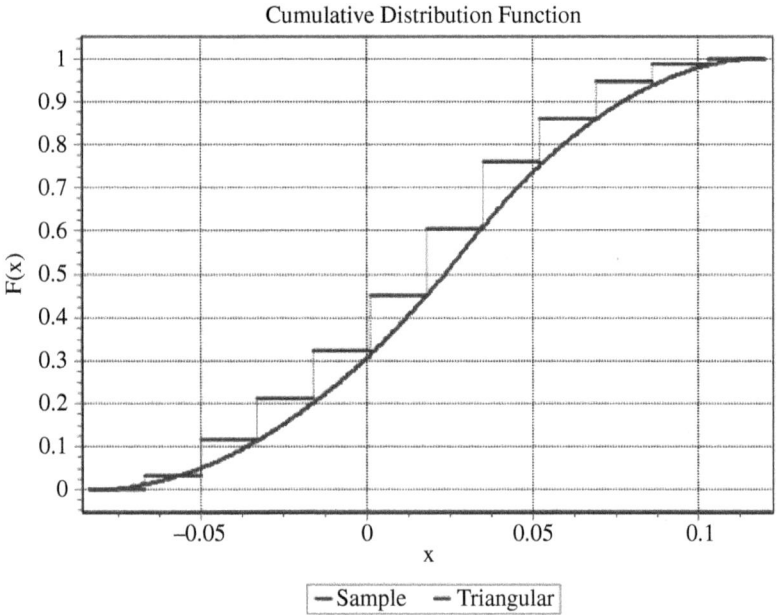

Figure 4.49. Cumulative distribution for BLS 1 in surge response (case 1).

are: $m = 0.029$; $a = -0.08449$; and $b = 0.12048$. CDF is computed as follows:

$$
\mathrm{CDF} = \begin{cases}
0 & \text{for } x \leq a, \\
\dfrac{(x-a)^2}{(b-a)(m-a)} & \text{for } a < x \leq m, \\
1 - \dfrac{(b-x)^2}{(b-a)(b-m)} & \text{for } m < x < b, \\
1 & \text{for } b \leq x.
\end{cases}
$$

In this case, the value of $\{x\}$ lies in between (m) and upper limit (b). Probability of survival is determined as 0.9886 and hence the probability of exceedance (failure) is 0.01278, which ensures that the surge displacement of the buoyant leg (BLS 1) does not exceed the threshold value even under the postulated failure case 1. This may be attributed to the fact that the eccentric load is acting on the same leg and hence surge acceleration does not exceed due to the eccentric load placed on the same point.

Similarly, surge displacement of the buoyant leg (BLS 1) under the postulated failure case 3 shows the Gumbel max probability distribution. As seen from the table, parameters of this distribution are: $\mu = -0.00568$ and $\sigma = 0.05924$. CDF is computed as follows:

$$F(x; \varphi, \beta) = e^{-e^{-(x-\varphi)/\beta}},$$

where $\varphi = \mu - \gamma\beta$. Here γ is Euler–Mascheroni constant $(= 0.5772)$ which results in the value of the variable $\{\varphi\}$ as -0.0323.

$$\sigma = \beta\pi/\sqrt{6}.$$

For the present case, it is computed as 0.04621. Subsequently, the probability of survival is 0.6297 and hence the probability of exceedance is 0.3702. This confirms that the surge response exceeds the threshold by about 14.3%. Similarly, the probability of exceedance is calculated for each postulated failure case and listed in Table 4.14. Reliability (R) is given by

$$R = 1 - P(f),$$

where $P(f)$ in this case is the probability of exceedance. Probability of exceedance, as seen from the table, is either too high or too low is some cases. Hence, monitoring displacement in all degrees of freedom alone will not be sufficient to detect the postulated failure, whereas this is only one of the factors in damage detection. In this study, the wireless SHM system is designed to measure the acceleration and rotations while the wired SHM system is deployed to measure axial loads. This can be useful to estimate the service life of tethers under cyclic loads through fatigue analysis.

4.39. Service Life of Tethers Under Postulated Failure

One of the most common techniques used to assess the reliability of components against fatigue failure is the S–N curve approach. When the platform is subjected to continuous wave loading, tethers undergo dynamic tether tension variations; this shall lead to fatigue failure. Fatigue is defined as the failure of a metal under repeated or varying load, which never reaches a level sufficient to cause failure in a single application (Pook, 1983). The service life of the tethers under postulated failure case 3 is estimated. Offshore compliant structures are generally designed to remain positive buoyant; hence,the overall failure of the structure is a very rare phenomenon. It is important to note that compliant offshore platforms do not completely fail even under the pull out of one of the tethers. This is due to the fact that the load will be distributed among successive tethers. But, under sequence of such failures, platform may not remain functional.

Fatigue design is based on the use of S–N curves, which are obtained from fatigue tests of standard specimens. While most of the non-ferrous metals do not exhibit endurance limit, fatigue strength is defined as the maximum cyclic stress range that can be applied without causing failure for a defined number of cycles; this value is normally 10^7 cycles. When plotted on a log–log scale, the S–N curve can be approximated by a straight line. A power law equation can then be used to define the S–N relationship, as follows:

$$NS^m = A$$
$$\Rightarrow \quad \log N = \log A - m \log S.$$

where N is number of cycles, A and m are constants obtained from the S–N curve and S is the stress range. For the present study, the S–N curve in sea water with cathodic protection is considered. Rain-flow counting algorithm is used to estimate the stress range from the histogram value. Number of counts for each stress bins is obtained from the rain-flow counting method. Subsequently, N is estimated from the equation. (Chandrasekaran, 2015a–c). Miner proposed a linear summation hypothesis according to which damage fractions that are caused by different range of cyclic stresses could be summed to assess the overall damage. This is given by

$$D = \sum_{i=1}^{m} \frac{n_i}{N_i},$$

where D is total damage, n is number of counts obtained from the histogram, N is the stress range and m is the number of stress bins.

Scaled model of BLSRP is subjected to extreme wave height of 16 cms to acquire dynamic tether tension variations using ring-type load cells. Stress time history is calculated for each tether. Using the rain-flow counting algorithm, stress bins and number of counts for each stress bin are calculated. Number of cycles is subsequently computed from the S–N curve (DNV-RP-C203). Using Miner's rule, the sum of damage for all stress ranges is computed. This value is then extrapolated to obtain a value of unity to derive the service life of tethers under the postulated failure case. A sample calculation of damage of one of the buoyant legs (BLS 5) is shown in Table 4.16. The total period of acquisition is 81 s for which the damage is estimated as

$$D = \sum_{i=1}^{m} \frac{n_i}{N_i} = 3.67177\text{E} - 03,$$

where m is the total number of stress bins. Damage is estimated as 1.003 $(= (3.67177\text{E} - 03 \times 60)/81 \times 60 \times 6.2)$. The estimated damage would be equivalent to 1 in 6.2 hours, which is considered as the service life of the structure under extreme sea state considered in the study. Similarly, the damage is estimated for all tethers under the postulated failure case 3 and the corresponding service life is

Table 4.16. Damage estimate of BLS 5 under postulated failure (case 3).

Stress centre	Counts	Cycles	Damage
230.5754	1	2419627.3	4.1329E−07
236.8236	0	2174203.2	0
243.0718	12	1959121.8	6.1252E−06
249.3201	18	1769990.7	1.017E−05
255.5683	80	1603141.8	4.9902E−05
261.8166	392	1455497.8	2.6932E−04
268.0648	304	1324466	2.2953E−04
274.313	294	1207853.2	2.4341E−04
280.5613	536	1103796.5	4.856E−04
286.8095	188	1010708	1.8601E−04
293.0578	308	927229.08	3.3217E−04
299.306	123	852193.56	1.4433E−04
305.5543	130	784597.06	1.6569E−04
311.8025	229	723571.85	3.1649E−04
318.0507	81	668366.06	1.2119E−04
324.299	64	618326.25	1.0351E−04
330.5472	133	572883.05	2.3216E−04
336.7955	53	531538.94	9.971E−05
343.0437	53	493858.12	1.0732E−04
349.2919	99	459457.91	2.1547E−04
355.5402	32	428001.44	7.4766E−05
361.7884	50	399191.53	1.2525E−04
368.0367	16	372765.44	4.2922E−05
374.2849	11	348490.33	3.1565E−05
380.5332	13	326159.48	3.9858E−05
386.7814	3	305589	9.8171E−06
393.0296	5	286614.93	1.7445E−05
399.2779	2	269090.9	7.4324E−06
405.5261	0	252885.94	0
411.7744	1	237882.67	4.2038E−06

given in Table 4.17. It is seen from Table 4.17 that the service life of the buoyant leg BLS 1 is very less due to the fact that it undergoes significant tether tension variations as it encounters waves at the entrant point.

Table 4.17. Service life of buoyant legs under postulated failure (case 3).

Buoyant leg	Service life (in hours)
1	1.0
2	3.6
3	4.45
4	5.35
5	6.2
6	2.0

4.40. Frequency-Domain Approach for Postulated Failure

Frequency-domain responses for various postulated failure cases of BLSRP are discussed in this section. Figure 4.50 shows the PSD of the deck response in surge degree of freedom during postulated failure cases 1 and 2 under regular waves (wave height = 10 cm; period 2 s). It is seen from the figures that the peak frequency occurring at 0.52 Hz under normal loading conditions shifts to 0.56 Hz under postulated failure case 1. Under the postulated failure case 2, as seen in the figure, two peaks are seen at 0.46 Hz and 0.92 Hz in addition to two minor peaks; one at a very low frequency and the other at 1.36 Hz, which is thrice that of the first peak frequency. While occurrences of smaller peaks are attributed to the second-order vibrations caused by tether failure, higher frequency responses of the deck show the influence of response of buoyant legs on the deck motion due to their stiff connectivity in the vertical plane. When one of the tethers is removed, as in case 2, the platform still remains floating but the deck response is magnified.

The PSD of the deck response in heave degree of freedom is shown in Figure 4.51 under the postulated failure case 2. It is seen from the figures that the first peak, which is occurring at about 0.5 Hz under normal loads, does not shift even under the postulated failure except with a significant increase in magnitude. Reduction in the magnitude of the second peak, occurring at about 1 Hz, shows the influence of the buoyant leg on the deck response in heave motion. Peaks occurring at higher frequencies are attributed

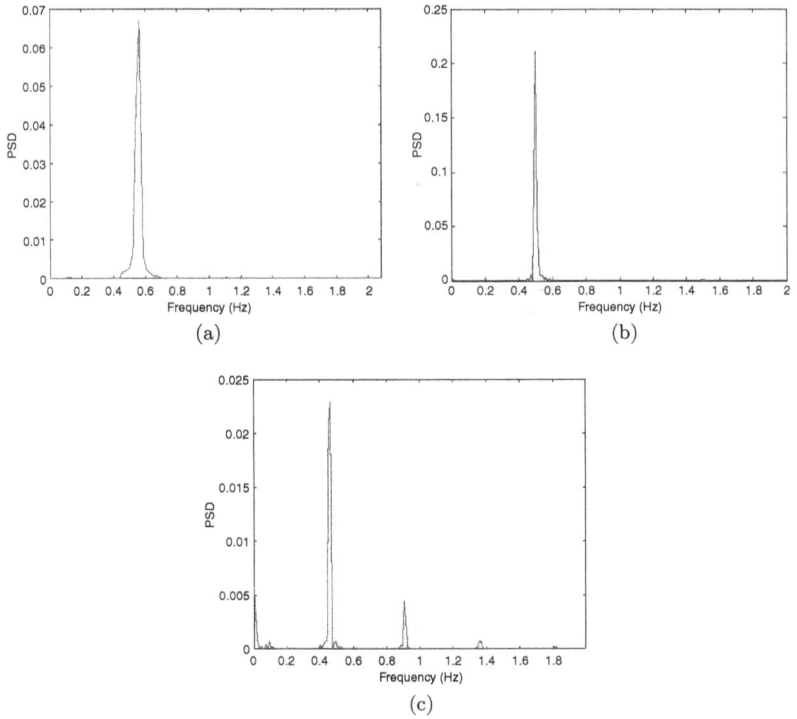

Figure 4.50. Surge response of deck under postulated failure (case 1): (a) normalcase; (b) eccentric load (case 1); and (c) tether pull out (case 2).

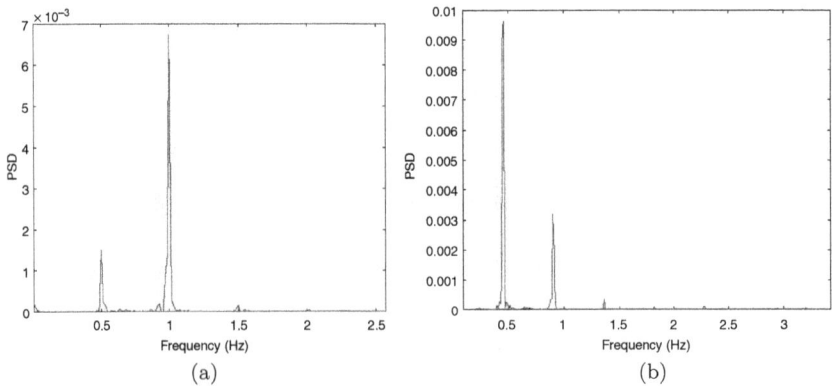

Figure 4.51. Heave response of deck under postulated failure (case 2): (a) normal case; and (b) tether pull out (damage case 2).

to the pitch motion of the deck, resulting from differential heave. Under the postulated failure of tether pull out, differential heave becomes predominant. This causes magnification of the first peak without any significant change in the frequency band. It is vital to note that the deployed SHM network is capable of capturing this behaviour satisfactorily. Smaller peaks seen at 2.5 and 3.5 Hz are not magnified and hence not transferred to the deck due to the isolation of superstructure from the base.

Figure 4.52 shows the PSD of heave response of one of the buoyant legs (BLS 4) under both normal case and postulated failure case 1. Where there is no postulated failure, the first peak occurs at 0.52 Hz, which shows the maximum amplitude. Other successive peaks show coupling effects of the response of buoyant legs on the response of the deck response. This is due to the fact that frequencies of occurrence correspond to those of the buoyant legs in surge and heave modes. Case 1 is due to the addition of eccentric load on BLS 1. Due to the eccentric load added to BLS 1, the heave magnitude value is smaller than the normal one. Variations in peak amplitude of heave response of BLS 4 are magnified due to the eccentric load added at BLS 1 location; it is important to note that both the legs, BLS 1 and BLS 4, are positioned diagonally opposite to each other. Even for a lower variation in the magnitude, response is seen in a

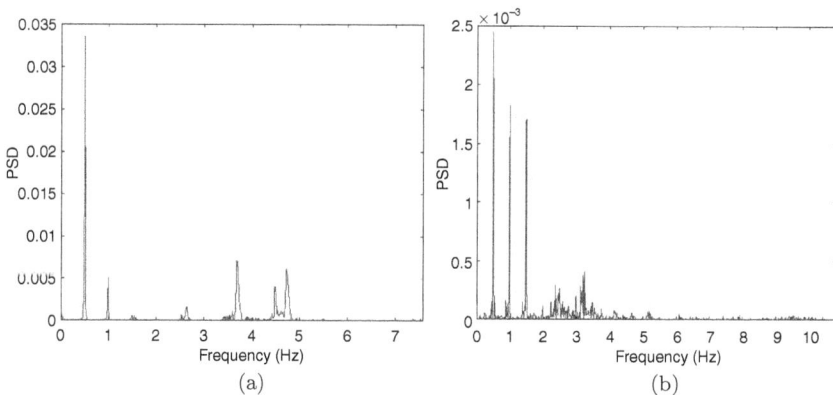

Figure 4.52. Heave response of BLS 4 under postulated failure (case 1): (a) normal case; and (b) eccentric load (case 1).

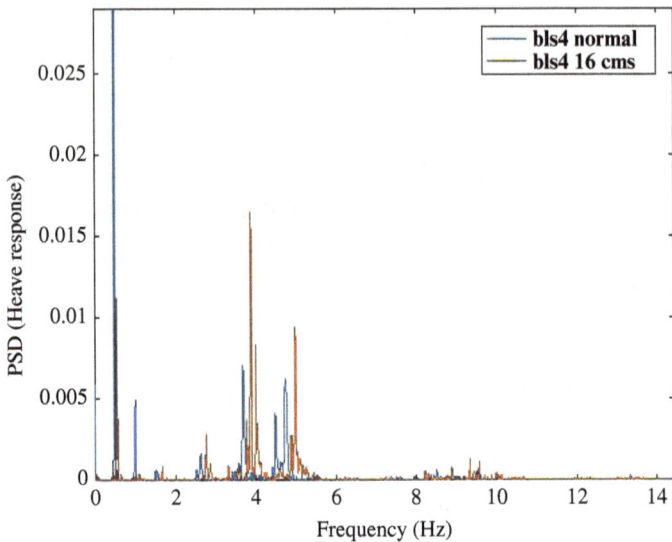

Figure 4.53. Comparison of heave response of BLS 4 under normal and extreme waves.

wide range of frequencies indicating an alarming behaviour of the structure. Response variations due to eccentric loading in BLS 1 is not uniform in all the buoyant legs. As BLS 4 is oriented opposite to that of BLS 1, addition of eccentric load on BLS 1 showed significant changes in the response of BLS 4. Comparison of heave response of BLS 4 both under normal and postulate failure case 3 is shown in Figure 4.53. As this value exceeds the threshold value, it will trigger the AMS. However, at lower frequencies, PSD magnitude values do not exceed the threshold values.

4.41. Localisation of Frequencies Under Postulated Failure Cases

With the PSD plots, only global features of the postulated damages are extracted without any localisation in the time domain. To get the local features of frequency across the time value, STFT is carried out; results are plotted for a chosen hamming window. These data are useful in diagnosing the condition of the platform for further assessment. With the STFT plots, a wide range of frequency variations occurring throughout the time history are successfully captured. Localisation of frequencies with respect to time is given by the STFT. Figure 4.54 shows the STFT of deck response in surge degree of freedom under normal loading and postulated failure of tether pull out. Figure 4.55 shows STFT of the deck response in heave degree of freedom under postulated failure (case 2). It is seen from the figures that there is not much variation in the plots of the normal case and damage case 2, except the occurrence of second peak at 0.91 Hz. Figure 4.53 shows the deck response in heave degree of freedom under the normal case and damage case 2. It is seen from the figures that the amplitude values of the deck response are higher in comparison to that of the normal case, apart from a wide frequency distribution.

Figure 4.56 shows the STFT of heave response of BLS 4 under normal case and damage case 1. Variations in amplitude under normal case, seen at 0.5, 1.1, and 1.5 Hz, are magnified under the postulated failure at all peak frequencies unlike in case of tether pull out. Figure 4.57 shows the surge response of BLS 1 under normal and extreme wave (case 3). As seen in the figures, plots show a wide range of frequency distributions. The magnitude of the peak at 0.58 Hz is also slightly higher than that of the normal case. A wide spectrum of frequency distribution under the postulated damage case shows the necessity of DAQ for a wide bandwidth to capsulate efficient alert monitoring of the platform. Large set of frequencies and the corresponding variations along the timescale are successfully captured by the WSN.

4.42. Alert Monitoring

Alert monitoring system (AMS) is an integral part of the proposed SHM system, designed to diagnose the acquired response of the platform under both normal and postulated failure cases. AMS will be useful to quantitatively monitor the platform for a potential damage as well as gradual change in the response. Processing of the indigenously designed AMS is shown in Figure 4.58. While it is designed to capture the data through sensor nodes continuously, it will also check the acquired values against the preset threshold limits.

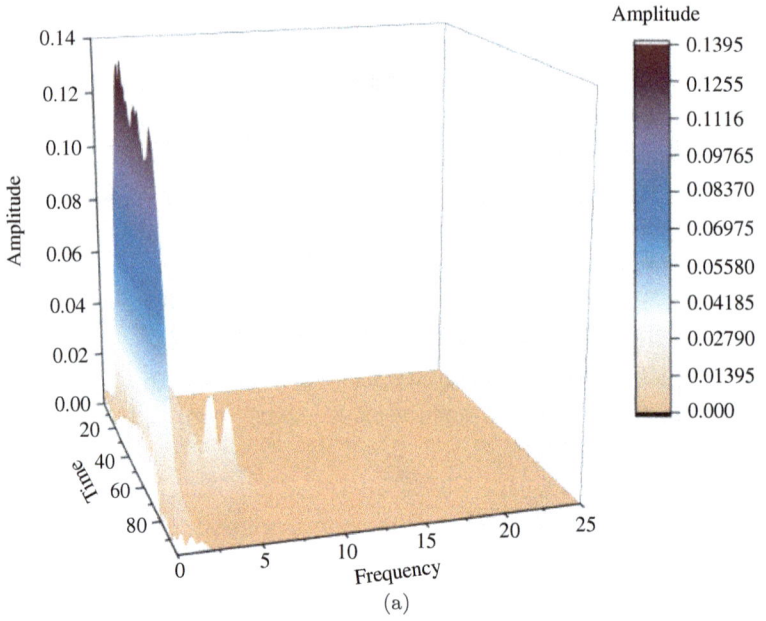

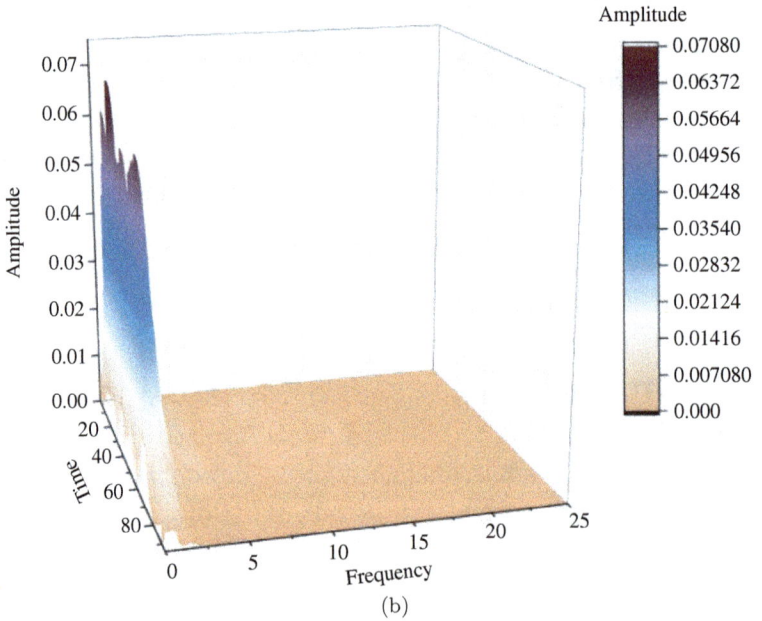

Figure 4.54. STFT of deck response in surge under postulated failure (case 2):
(a) normal case; and (b) postulated failure (case 2).

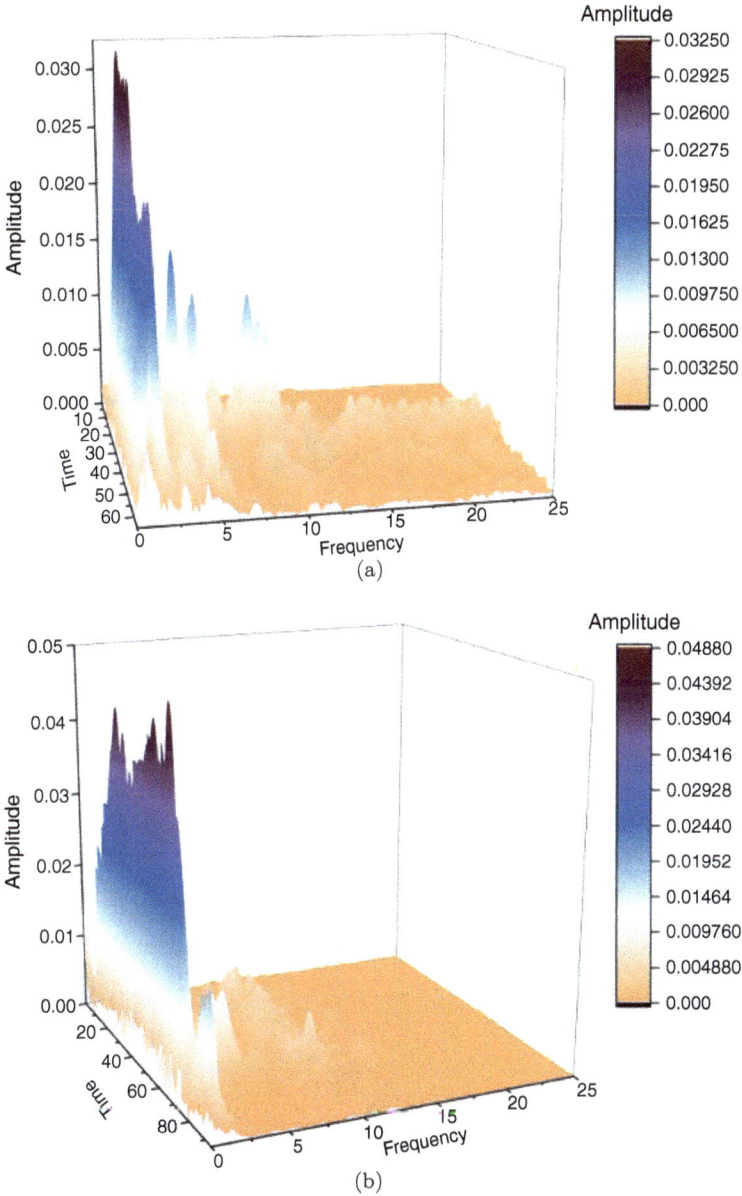

Figure 4.55. STFT of deck response in heave under postulated failure (case 2): (a) normal case; and (b) postulated failure (case 2).

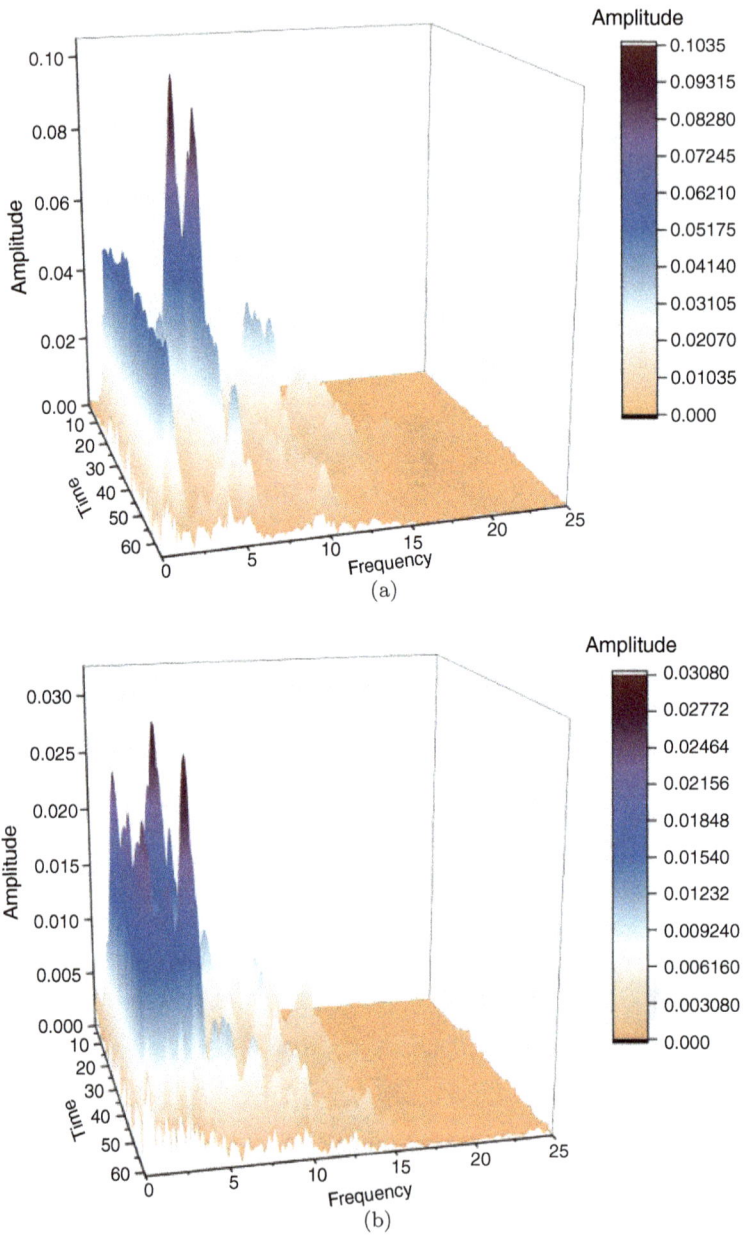

Figure 4.56. STFT of heave response of BLS 4 under postulated failure (case 1): (a) normal case; and (b) postulated failure (case 1).

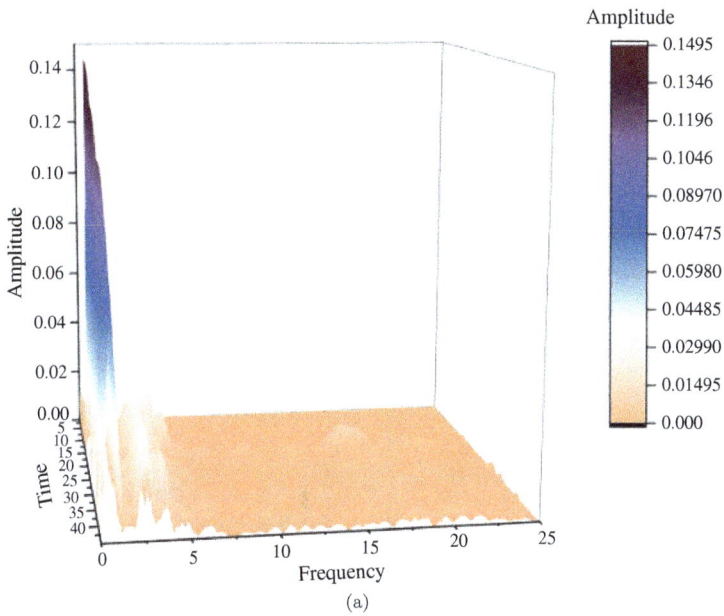

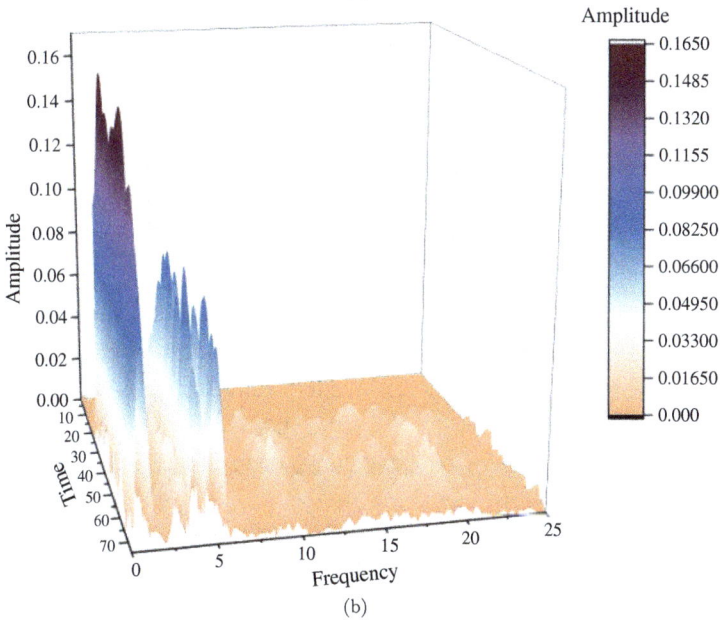

Figure 4.57. STFT of surge response of BLS 1 under postulated failure (case 3): (a) normal case; and (b) postulated failure (case 3).

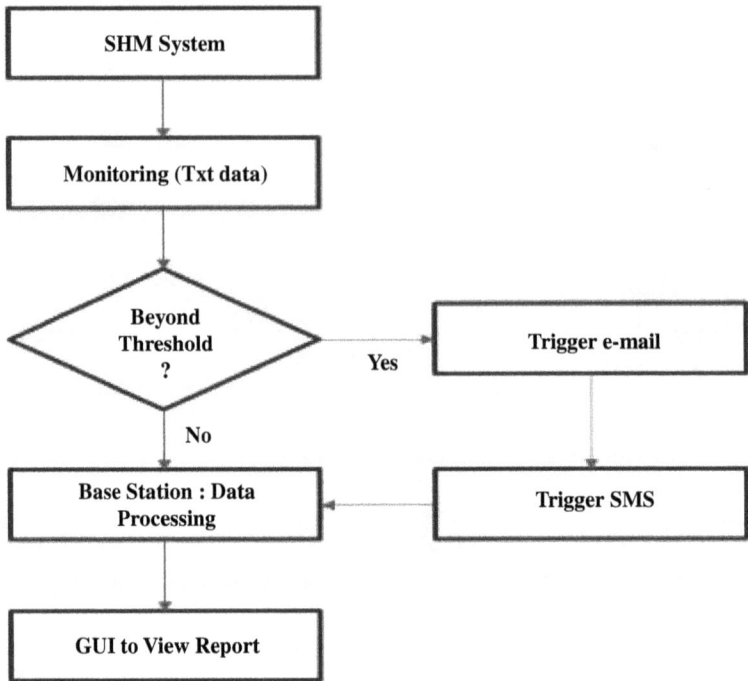

Figure 4.58. Algorithm of AMS.

On exceedance, it will trigger appropriate alert messages and send alert emails to the authenticated users. Alert messages with the details of the sensor node will also be displayed in the admin PC located at the base station. Admin can then check the time and details of the alert messages. In addition, authorised users shall can also access details of time and amplitude values of exceedance by clicking on the quick view tab of a particular sensor. Alert messages and emails, generated as an integral part of the AMS, will also display relevant details: (i) code number of the sensor; and (ii) sensor mapping in terms of its location.

4.42.1. *Identifying alarm events*

An alarm event is determined by examining the peaks of the acquired acceleration response. Alarm events may occur either due to the damage of members or under exceedance of response due to other reasons. Alarm events can also be useful to know whether the damage of the member is caused due to any shock or impact loads. Damage due to impact loading can be identified by observing the time stamp. A sudden peak occurring between the start and peak indices is a measure of caution of the presence of impact load; analysing the difference in their time stamp will confirm the presence of impact load. The alarm event will trigger necessary alert messages with details of sensor location, time stamp and exceedance value. It enables the user or admin in charge to be proactive in such cases. The proposed SHM along with the AMS is very useful in monitoring unmanned platforms in terms of its structural safety and security threats as well. Recent advancements in neural network and artificial intelligence shall support the decision-making process in integration with the SHM system. Versatility of the proposed SHM system is high due to the fact that sensor nodes can also be trained on the pattern of exceedance of threshold values. Once successful training is imparted to the sensor nodes based on the previous data, even preventive maintenance can become very efficient apart from improving operational and structural safety of the platform. Offshore accidents, though not very common, are highly catastrophic. Under the recent guidelines of HSE practices, the proposed SHM system is highly beneficial to take important decisions under envisaged damage scenarios.

4.43. Data Acquisition and Storage

In this study, only acceleration and displacements are considered as main parameters, for which slow-speed monitoring strategy is suitable. But to monitor large amplitude vibrations and responses of impulsive nature, a high-speed DAQ is preferred. Hence, data sampling rate should be chosen by the user. For long-term monitoring, datasets that are collected for extended period of time are considered to be the main issue. This is due to the fact that increase in size of storage is necessary. Monitoring offshore platforms continuously for a long period will result in a huge volume of data, which need to be subsequently processed and stored in the server.

In such cases, significant attention is given to incorporate data reduction and processing algorithms by retaining the useful information.

4.44.   User Interface and Report Generation

Data acquired by the sensor nodes are transmitted to the base station and stored in a MySQL database. Acquired data are processed and enabled for viewing as a report at the user interface. User interface can be accessed both by the server at the base station and the authenticated user through the web. User interface is designed to contain information with respect to nomenclature of sensor nodes, a manual for sensor mapping to show its physical location, etc. in the real-time environment. In addition, users can also access more specific details of the response at any particular time window by specifying the period of interest. By default, the proposed user interface shows details of the acquired time history data for the last one hour; this can be displayed by clicking on the sensor data. If the user needs more details of previous records, one can access them by choosing the required period. Figure 4.59 shows the screenshot of the proposed user interface and Figure 4.60 shows detailed reports of each sensor node, which can also be viewed along with a plot for all degrees of freedom.

User interface is designed to generate reports for each of the sensors in the network. By clicking the specific sensor, it will generate the report along with the time stamp. In addition to this, a graph is also plotted for the last hour data. Software is developed to provide the following information:

(i) detailed report for each sensor;
(ii) graph plotted for all the degrees of freedom for the last 1 hour data;
(iii) complete list of values exceeding the threshold in the alert view.

Each report will contain the following information: (i) monitoring period during which outputs are given; (ii) graphical representation; and (iii) details of values exceeding the threshold. There is an alert and quick view tab for each sensor in the home page. By clicking on

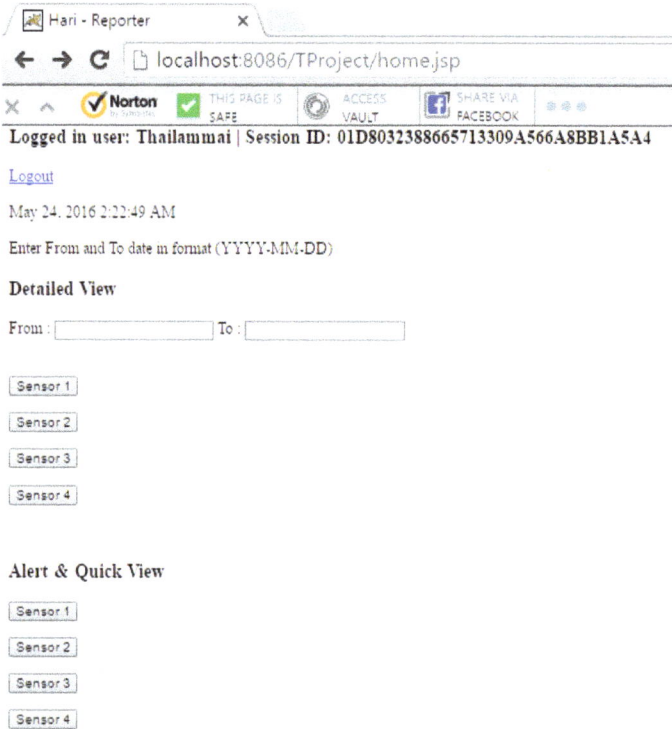

Figure 4.59. Home page of user interface.

the quick view tab of any particular sensor, only exceeded values will be displayed. This page also gives a quick summary of the minimum and maximum values in all degrees of freedom. Figure 4.61 shows the alert and quick view of the sensor while Figure 4.62 shows details of a typical sensor in quick view page.

Acceptable limit of the acceleration value is fixed as a threshold value, which can be changed in the user interface menu. On exceedance, alert messages will be displayed at the user interface. As sensor nodes and server units are connected to the same network, on exceedance of threshold value, the processor unit in the sensor node will trigger the email instantaneously. In addition, an appropriate alert message will also be triggered along with the sensor node details.

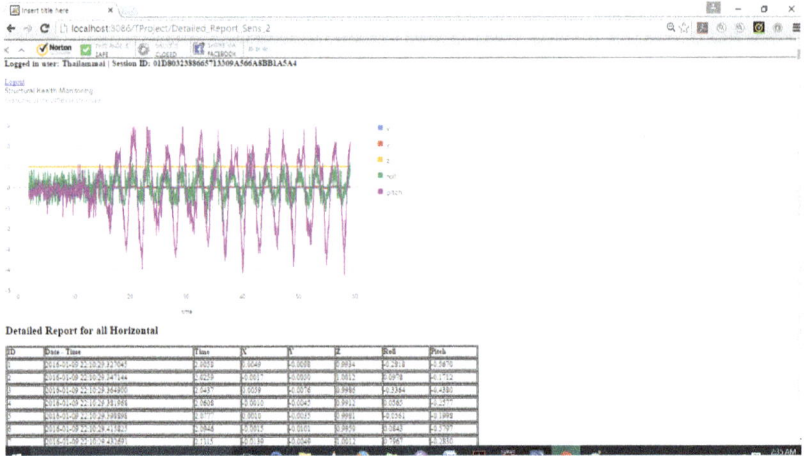

Figure 4.60. Detailed view of the sensor node.

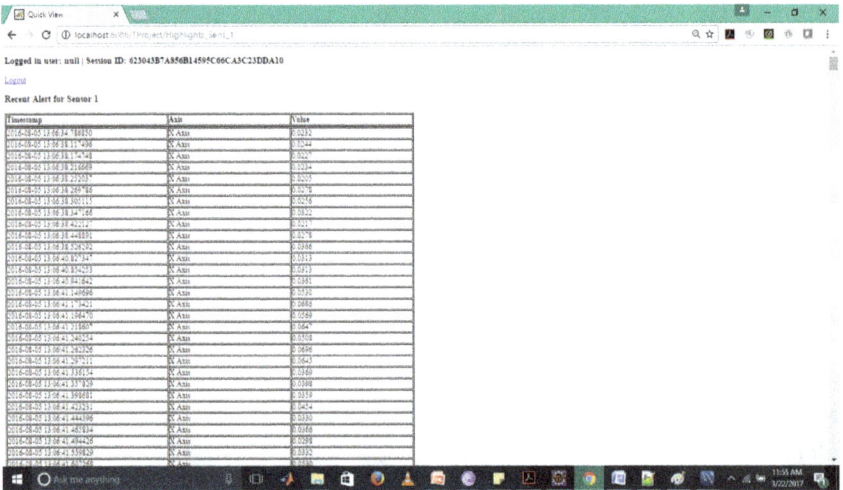

Figure 4.61. Alert and quick view of sensor.

Hence, location of the damage can be easily identified by mapping the sensor node location. After processing the data, the central server will also send SMS to the registered mobile number on exceedance of the threshold value. With the SMS API, SMS is triggered from the website for two-factor authentication.

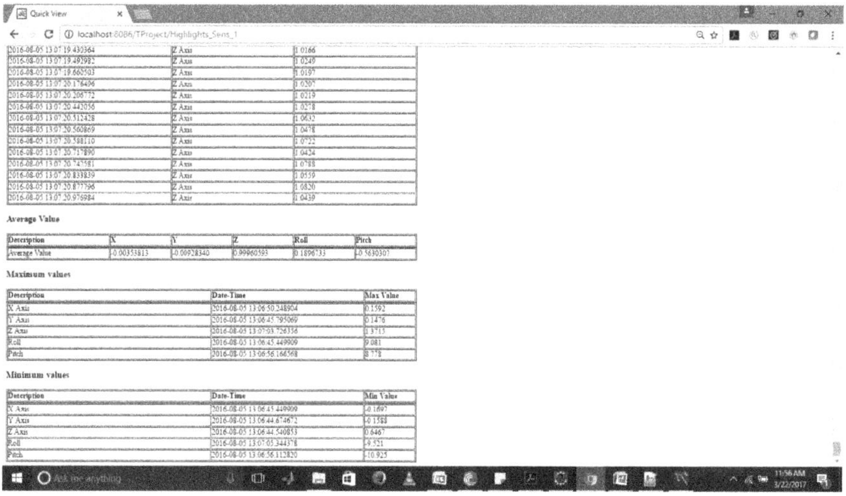

Figure 4.62. Values of the sensor in quick view page.

4.45. Threshold Value

Scale model is excited by a regular wave of 16 cm wave height (corresponds to very severe sea state in prototype) to assess the efficiency of the proposed AMS. Comparison of RAO plots for surge response is shown in Figure 4.63. As seen from the figures, the acceleration value exceeds the normal case at all time periods. Considering the normal case without any damage as the reference signal, the maximum amplitude is taken as the threshold value for the particular case. However, limit values of response in the active degrees of freedom are user-defined and can be fixed based on the previous time history records.

4.45.1. *Alert messages*

Under the postulated failure case 3 (extreme waves), the response of the platform exceeds the threshold value in surge, pitch, and heave degrees of freedom. Alert event triggers the alert messages in three modes: (i) alert messages displayed at user interface; (ii) alert email triggered to the registered user; and (iii) alert SMS triggered to the authenticated user. Alert message is triggered and displayed in the user interface when it exceeds the value. The alert message will

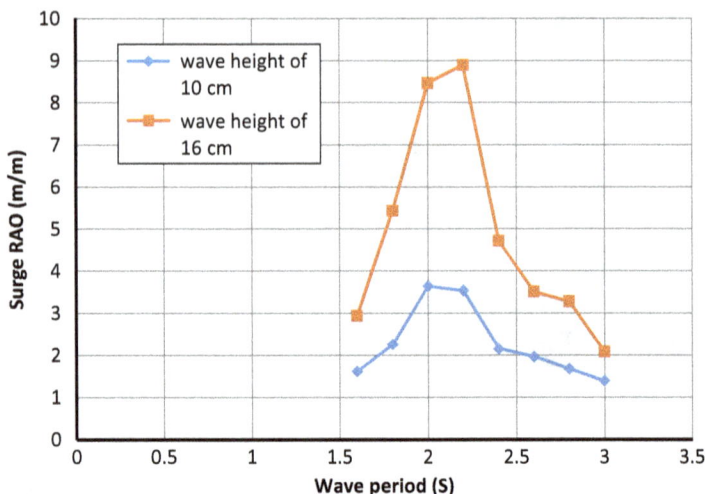

Figure 4.63. Comparison of surge RAO of deck for threshold values.

contain relevant information about the critical value and the details of sensors. Detailed view of each sensor can be viewed by clicking the corresponding sensor on the graph shown in the user interface. Complete details of the value, exceeding threshold are obtained by clicking the alert view. When the threshold value is exceeded, an alert message is triggered. The proposed SHM system shall first trigger an alert email to the registered user. Subsequently, data are transmitted to the base station; after further processing, SMS alert will be triggered by the web server.

4.46.   Discrepancy Between Lab Scale and Real-Time Monitoring

Although the proposed SHM is successfully demonstrated in lab scale, few factors need to be considered while implementing this design in real-time monitoring. While the layout of the sensor network and interfacing of hardware components remain unaltered for both the lab scale and real-time monitoring, bandwidth and latency issues are vital while implementing the software for the real-time monitoring. In addition, hardware configurations of the sensor units should be chosen based on the data requirements. Further, IEEE 802.11 protocol used in lab scale is not compatible with all kinds of ocean environments; alternates include voice over IP, broadband data and video communication services for different topologies of sensors. Improvements in satellite, VSATs and antenna systems will enable it to meet the demand for higher bandwidth in offshore platforms. Scalability is also an important issue in real-time monitoring. Cluster topology and multi-hop communication are required for scalability of low power WSNs but at the cost of more complex routing protocols. Without major modifications in the proposed WSN, modified protocols can be used to handle scalability in real-time monitoring.

4.47. Seismic Assessment of a Bus Duct for SHM

A 6.6-kV segregated phase bus duct, which is primarily used in power transmission of a **XXX** Nuclear Power Plant is assessed for its seismic safety; location of the plant is masked for strategic reasons. Acceleration spectrum conforming to Safe Shutdown Earthquake with 4% damping is considered as the input load for the seismic analysis. Modes of vibration of the bus duct are reexamined for any noticeable modes in torsion, as fundamental modes. Connectors, which are used to suspend the bus duct from the roof slab, are examined for axial stresses in the anchor bolts. Postulated failure of one or two supports randomly and its influence on the failure of the bus duct system is analysed for its safety against collapse prevention. Numerical analyses are found helpful in improving the design of the bus duct structure and its support system with an increased factor of safety.

With the present advancements in numerical analyses, most types of structures with complex geometry are analysed for their critical performance. Modal Frequency Response Analysis approach that utilises the mode shapes of the dynamic system is one of the common methods of such analyses (Chandrasekaran and Roy, 2006). While it reduces the size of the equations of motion by de-coupling them, it also makes the numerical solution more efficient

(Chandrasekaran and Roy, 2004; Chandrasekaran and Tripati, 2004). As mode shapes are typically computed as part of the dynamic characterisation of the structure, the modal frequency response is an extension of a normal modal analysis (Chandrasekaran and Tripati, 2005; Chandrasekaran *et al.*, 2004). Seismic analysis of structures of high importance has gained more focus after the recent damages caused to nuclear reactors in Japan; the damage caused by these loads on the structures is colossal (Spignesi, 2005). Checking for seismic safety is an inherent part of the structural design for nuclear reactors that are located in regions where seismic loadings are prevalent (Chandrasekaran *et al.*, 2006–2007; Reitherman, 1997; Wilson and Clough, 1999). Lateral displacements caused to structures require a detailed insight to reassess their functionality post earthquakes (Hudson, 1979, 1990; Chandrasekaran *et al.*, 2009). Recent studies show that most of the structures have higher modes of responses, which are uniquely activated during earthquakes (Chandrasekaran *et al.*, 2010). It is also highlighted that most damages are caused during the first few modes of vibration (Bozorgnia, 2003; Bozorgnia and Campbell, 2004; Campbell and Bozorgnia, 2003; Chandrasekaran *et al.*, 2007, 2008; Vamvatsikos and Cornell, 2002). Seismic analysis is effective to examine the complexities in the geometric designs of the structure with a special focus to ascertain the use of special earthquake-resistant elements (Chandrasekaran *et al.*, 2011). This example study deals with determining different modes of the bus duct to assess its seismic safety in the design perspective under a predefined earthquake spectrum that is applicable to nuclear reactors. Due to the geometric complexities involved in the bus duct design considered for the study, finite element analysis is preferred as it is capable of modelling the physical behaviour at very close time intervals (Joao, 2003).

4.47.1. *Numerical modelling of bus duct*

Present work shows the seismic response analysis of a 6.6 kV segregated phase bus duct of a Nuclear Reactor. Figure 4.64 shows the geometric configurations of the bus duct, which is suspended from

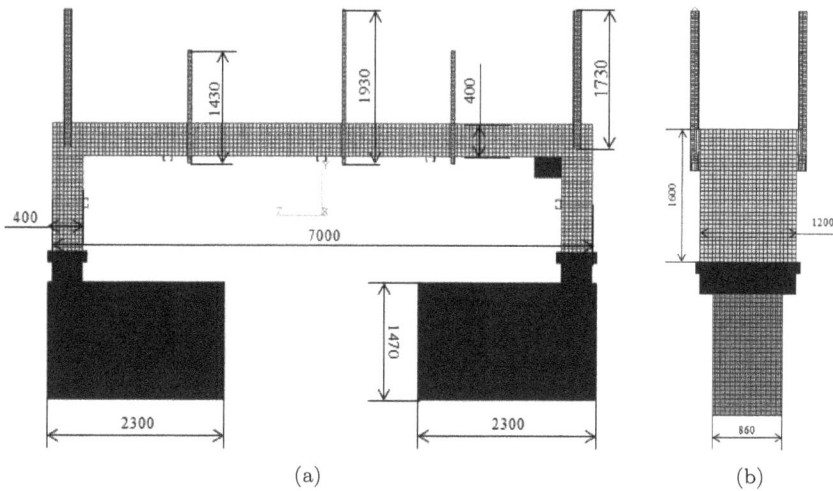

Figure 4.64. Model of bus duct. (a) Front view of the model of the bus duct; and (b) side view of the bus duct model.

Table 4.18. Material properties of bus duct.

Material	Young's modulus E, (Pa)	Density (kg/m³)	Poisson's ratio	Allowable stress (MPa)
Steel	2.0E11	7800	0.3	150
Aluminium	0.7E11	2700	0.3	80
Epoxy	1.08E10	981	0.32	—

the roof slab using K1- and K2-type connectors, which are ISMC 100 channels. Top ends of the K1- and K2-type connectors are assumed to be rigidly fixed to the slab. Upper part of the bus duct is made of 3 mm aluminum sheet and supported by ISMC 100 channels, which are insulated with epoxy insulators of 130 mm long and an outer diameter of 75 mm. Table 4.18 shows the structural details of the bus duct used in the present example study. User-specified floor acceleration spectrum is considered as the input for analyzing the bus duct; signal is applied at the top ends of the K1- and K2-type connectors in the y-direction. Figure 4.65 shows the input spectrum considered for the analysis. Figure 4.66 shows the numerical model

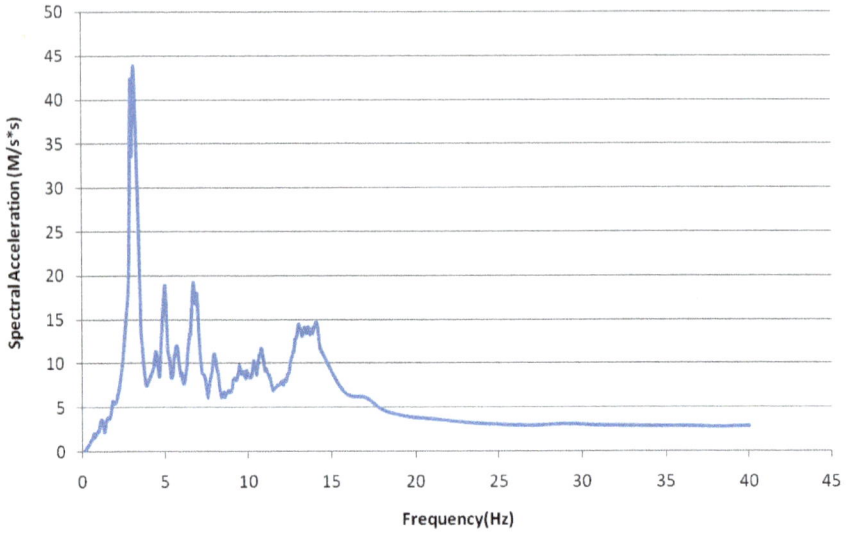

Figure 4.65. Floor acceleration spectrum at slab level along y-axis.

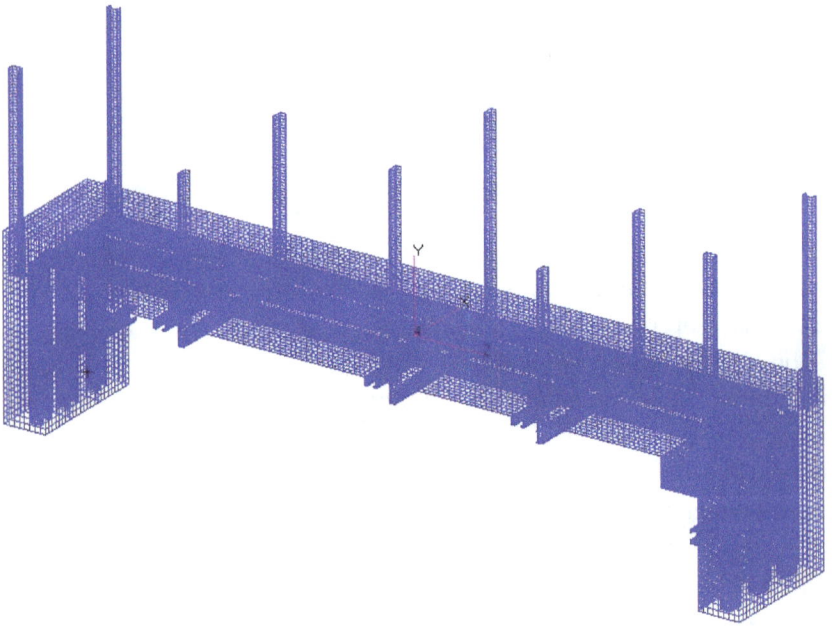

Figure 4.66. Numerical model of bus duct with connectors.

Table 4.19. Natural frequencies and mode description.

Mode no.	Frequency (Hz)	Description of the mode
1	5.155	Translational, Z-axis
2	7.496	Translational, X-axis
3	9.397	Translational, X-axis
4	13.726	Translational, X-axis
5	13.732	Translational, X-axis
6	13.751	Translational, X-axis
7	16.563	Translational, X-axis/slight rotation at corners
8	16.867	Translational, Z-axis
9	17.118	Translational, Z-axis
10	17.330	Translational, Z-axis
11	18.360	Translational, X-axis/slight rotation at corners
12	19.393	Translational, X-axis/slight rotation at corners
13	19.435	Translational, X-axis/slight rotation at corners
14	20.474	Translational, X-axis/rotation of the hanging parts
15	37.324	—
16	37.442	—
17	38.710	—
18	40.625	—
19	40.789	—
20	41.326	—

of the bust duct, K1- and K2-type connectors and details of the insulator. Table 4.19 shows the natural frequencies along with the description of the modes of the bus duct for the first few modes. It is seen from the table that more of the fundamental modes of vibration are translational. As no torsion modes are seen as fundamental modes, it reiterates the fact that the structural geometry of the bus duct is stable. Figures 4.67 and 4.68 show the first and second modes of vibration of the bus duct, respectively. Critical nodes are mapped from the finite element model and represented in the figures; location of these nodes can be easily visualised.

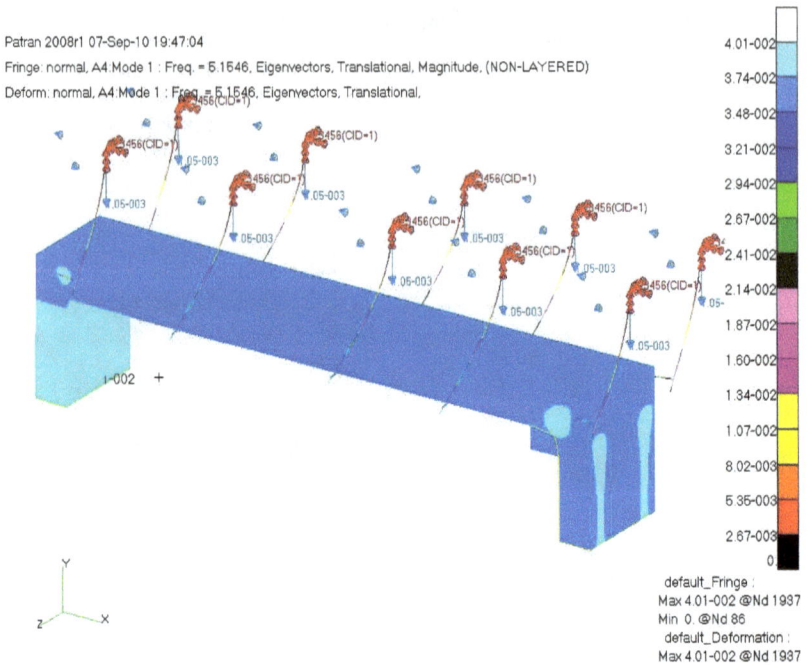

Figure 4.67. First mode of the bus duct (frequency = 5.154 Hz).

4.48. Structural Assessment of the Bus Duct

The seismic behaviour of bus duct, suspended by two different types of connectors is assessed. Connectors K1 and K2 are different in their geometric configurations. Results of the analysis is focused on specific nodes at which accelerations, displacement and von Mises equivalent stresses are shown. Nodes are chosen at the connection points of the bus duct and the connectors so as to insight the stress variations. Figure 4.69 shows the acceleration response spectrum at node A, which is the critical node on K1 connector while Figure 4.70 shows the displacement response spectrum at node A. Figure 4.71 shows the von Mises stress response spectrum at node A. It is seen from the figures that the peak acceleration responses occur at 2.85,7 and 19 Hz; near resonating-type displacement of amplitude 4.5 mm is seen at frequency 2.85 Hz. von Mises equivalent stress is about 27 MPa, which is much within the limits of allowable stress of Aluminum

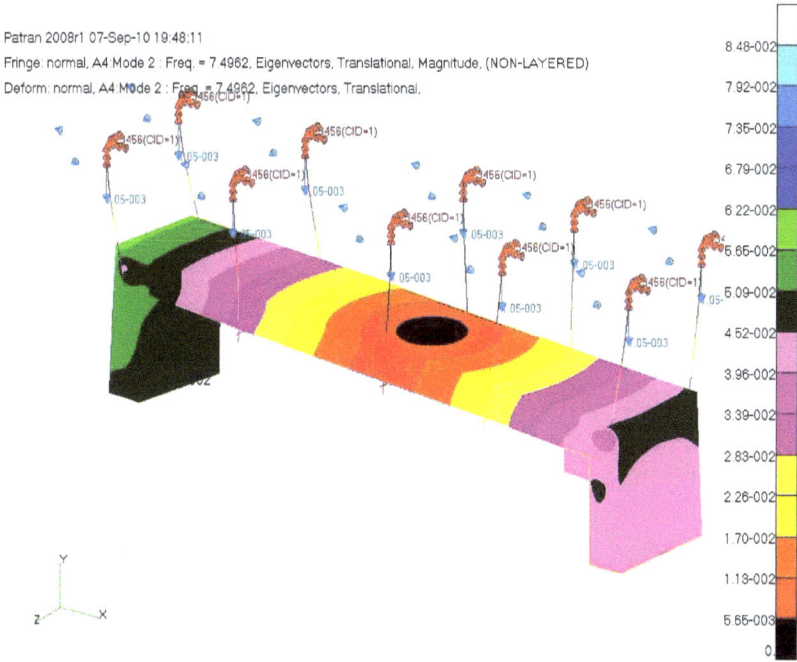

Figure 4.68. Second mode of the bus duct (frequency = 7.496 Hz).

(i.e. 80 MPa). This shows that the design of K1 connector is safe against the applied seismic loads.

Figure 4.72 shows the acceleration response spectrum at node B, which is the critical node on K2 connector while Figure 4.73 shows the displacement response spectrum at node B. It is seen from the figures that the peak acceleration response of 1.5 m/s^2 and near resonating-type displacement of amplitude 0.75 mm occur at 7 Hz. Similarly, Figures 4.74 and 4.75 show the acceleration and displacement response spectra at node C, which is the critical node on the short K2 connector. Peak acceleration response and near resonating displacement of amplitude are 1.4 m/s^2 and 3.6 mm, respectively. It is seen that they are much away from the resonance bandwidth of the bus duct, making it safe against the applied seismic loads. Figures 4.76 and 4.77 show the acceleration, displacement and von Mises stress spectra at node D, which is the critical node

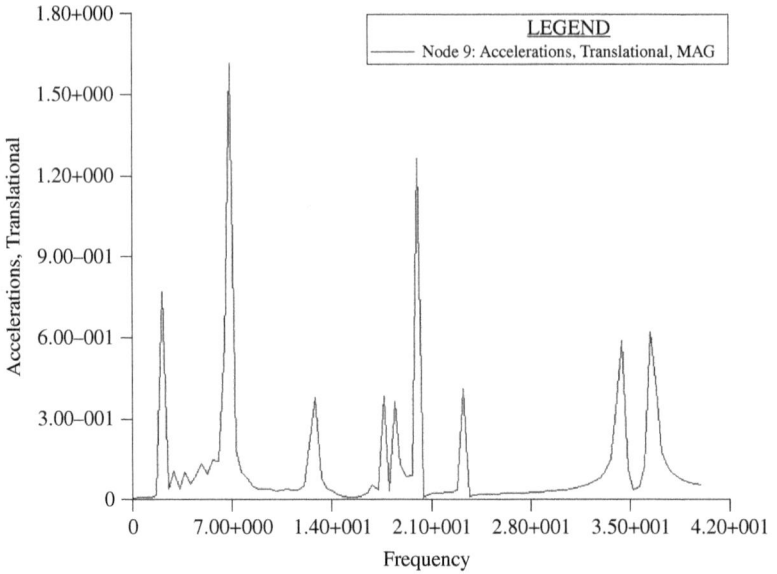

Figure 4.69. Acceleration response spectrum at node A (K1 support).

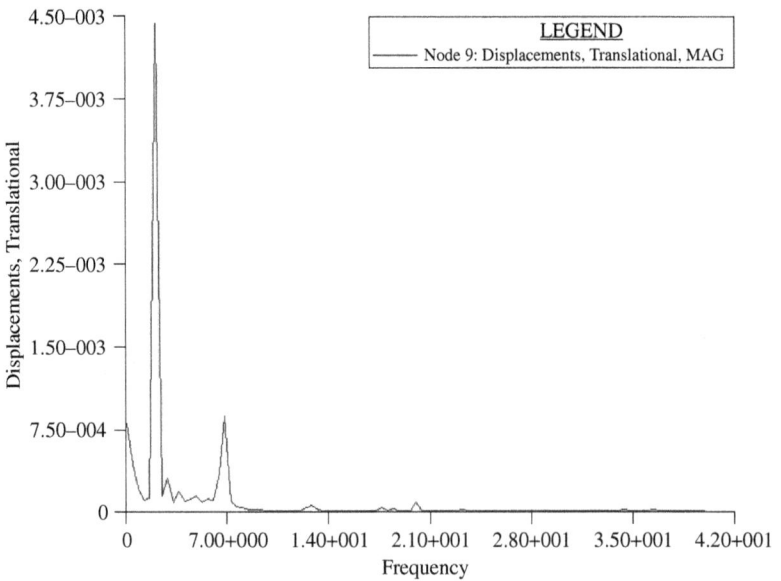

Figure 4.70. Displacement response spectrum at node A (K1 support).

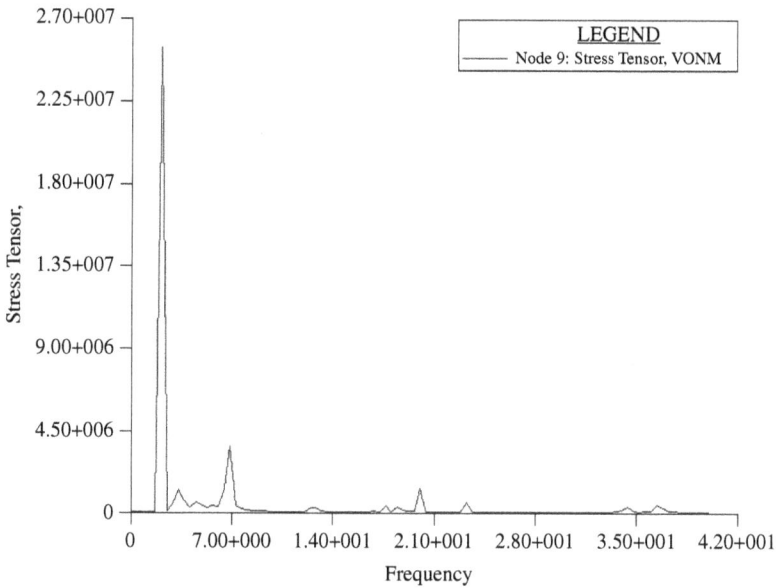

Figure 4.71. von Mises equivalent stress response spectrum at node A (K1 support).

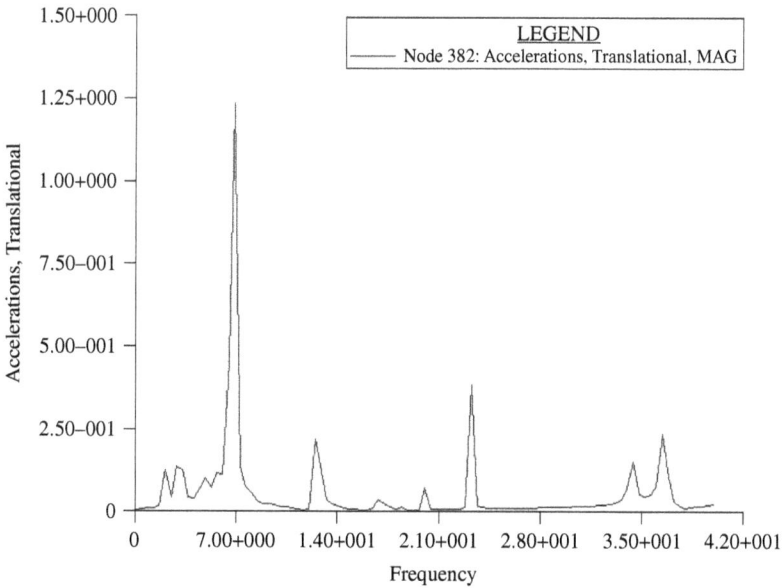

Figure 4.72. Acceleration response spectrum at node B (K2 long connector).

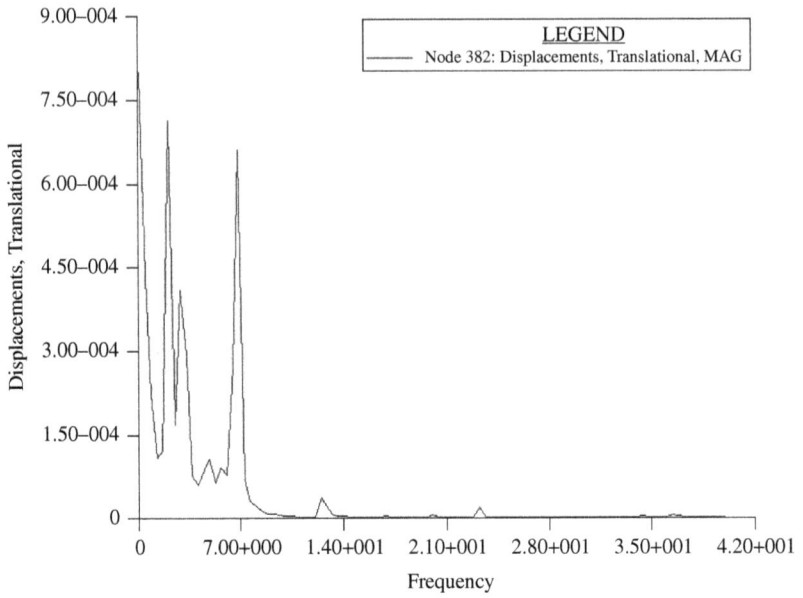

Figure 4.73. Displacement response spectrum at node B (K2 long connector).

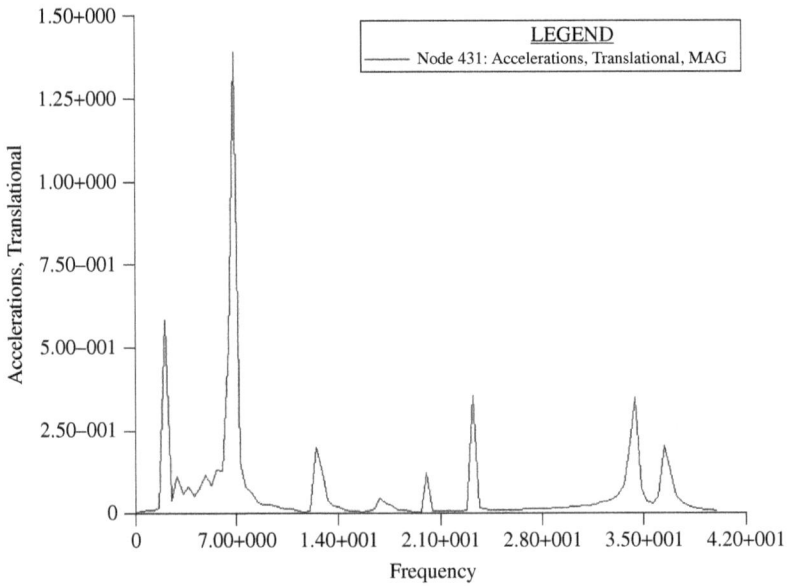

Figure 4.74. Acceleration response spectrum at node C (K2 short connector).

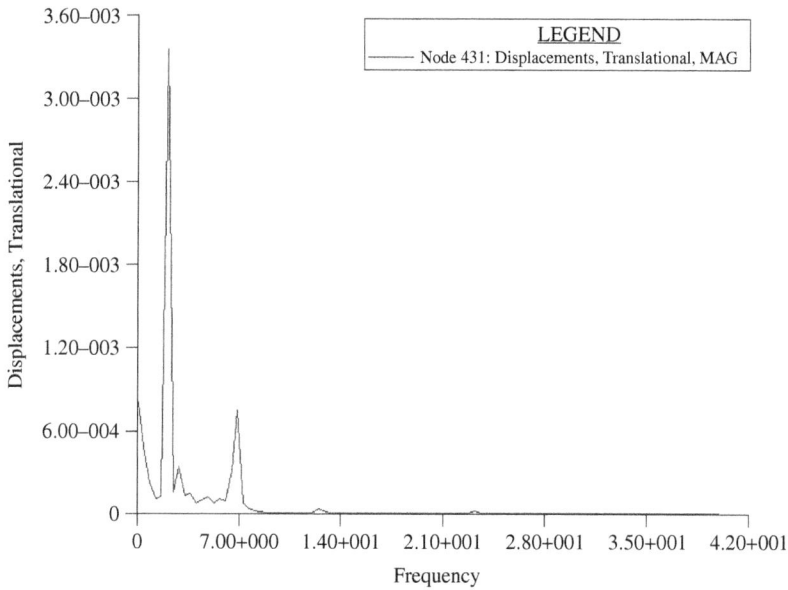

Figure 4.75. Displacement response spectrum at node C (K2 short connector).

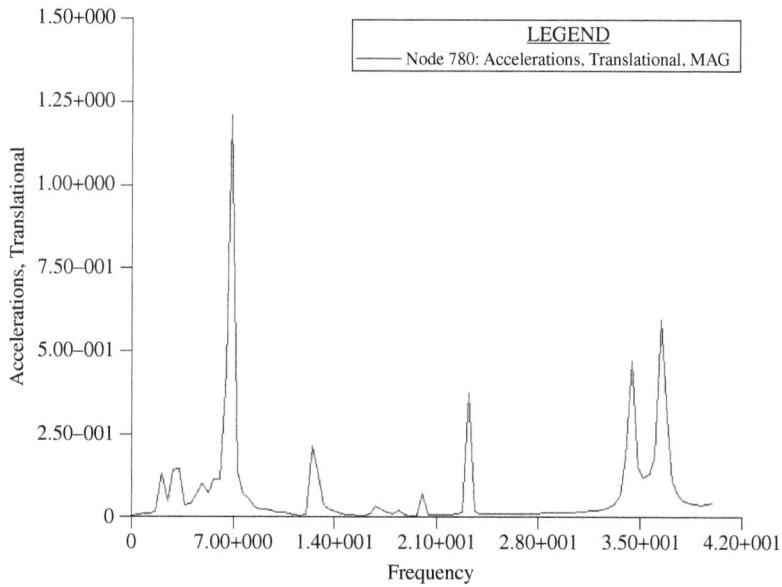

Figure 4.76. Acceleration response spectrum at node D (Insulator point).

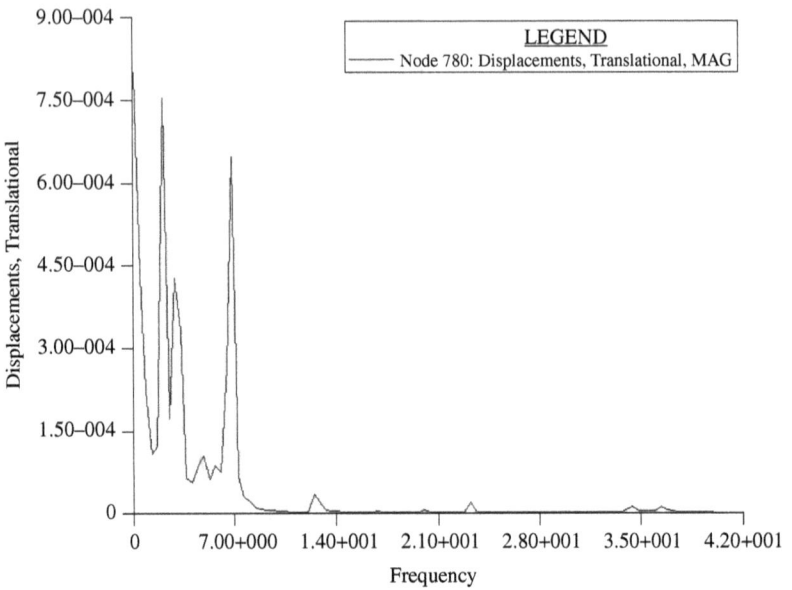

Figure 4.77. Displacement response spectrum at node D (Insulator point).

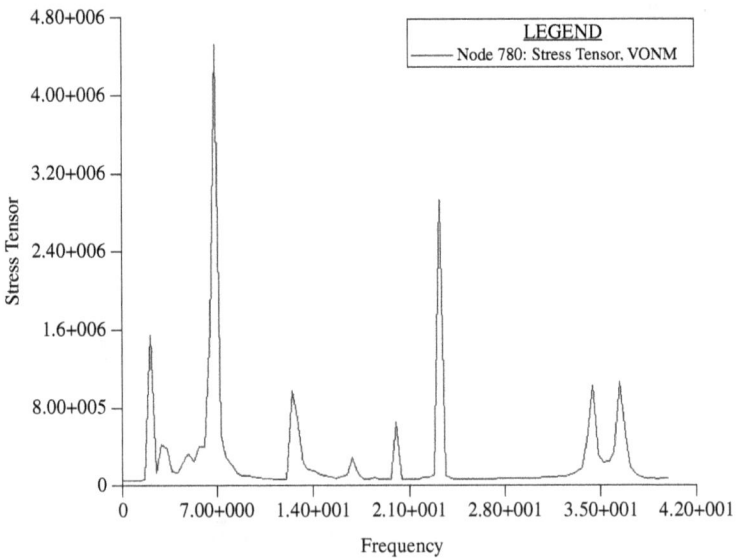

Figure 4.78. von Mises equivalent stress response spectrum at node D (Insulator point).

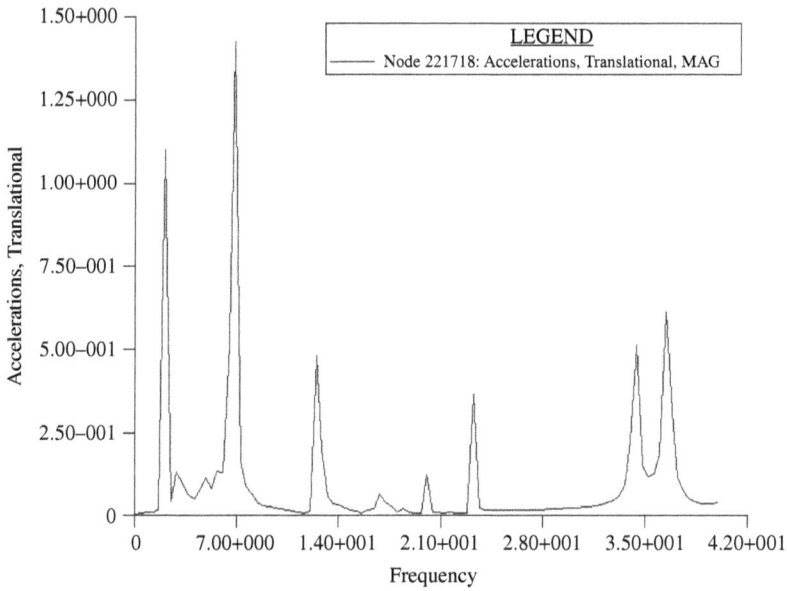

Figure 4.79. Acceleration response spectrum at node E (channel).

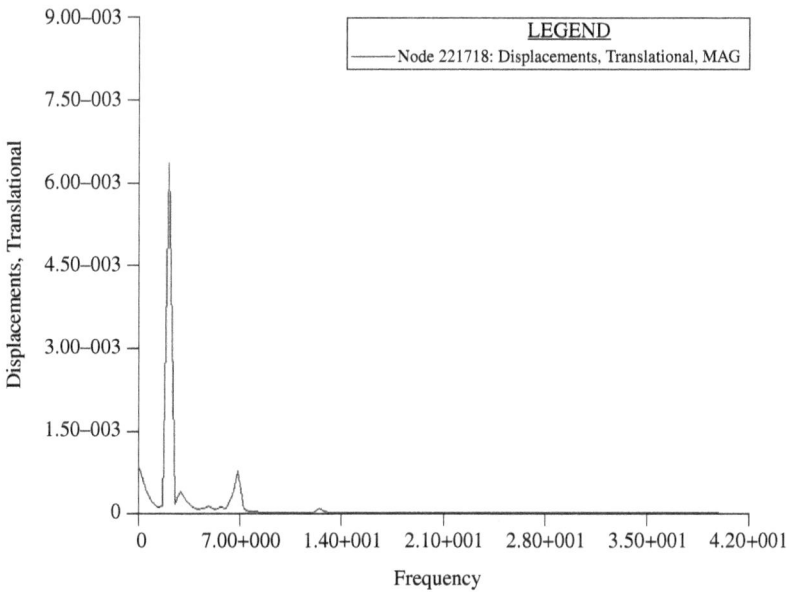

Figure 4.80. Displacement response spectrum at node E (channel).

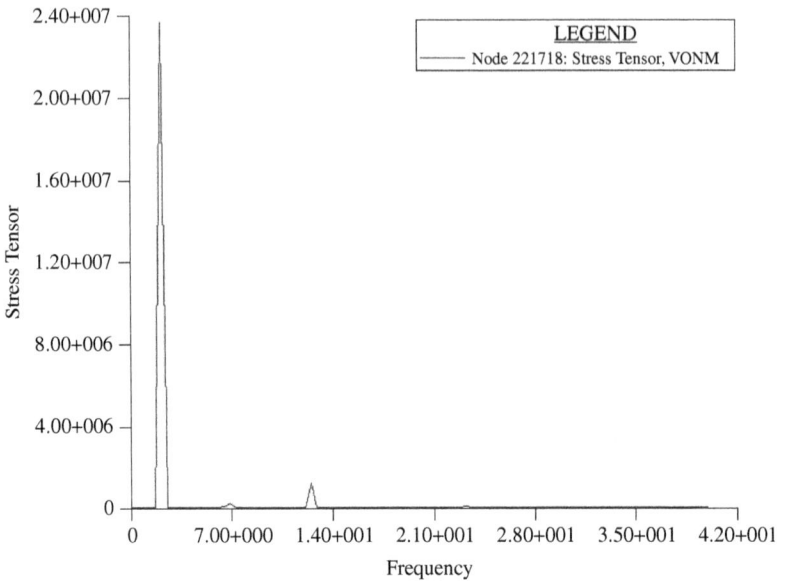

Figure 4.81. von Mises equivalent stress response spectrum at node E (channel).

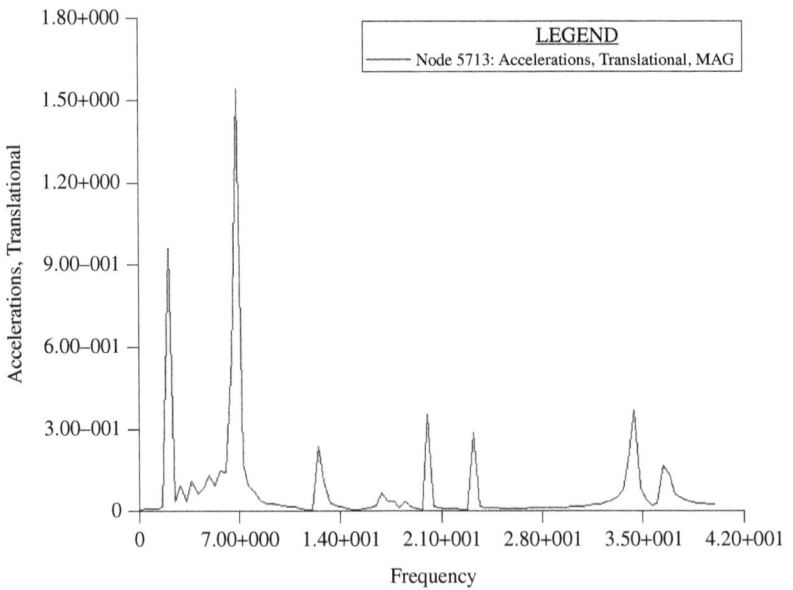

Figure 4.82. Acceleration response spectrum at node F (enclosure).

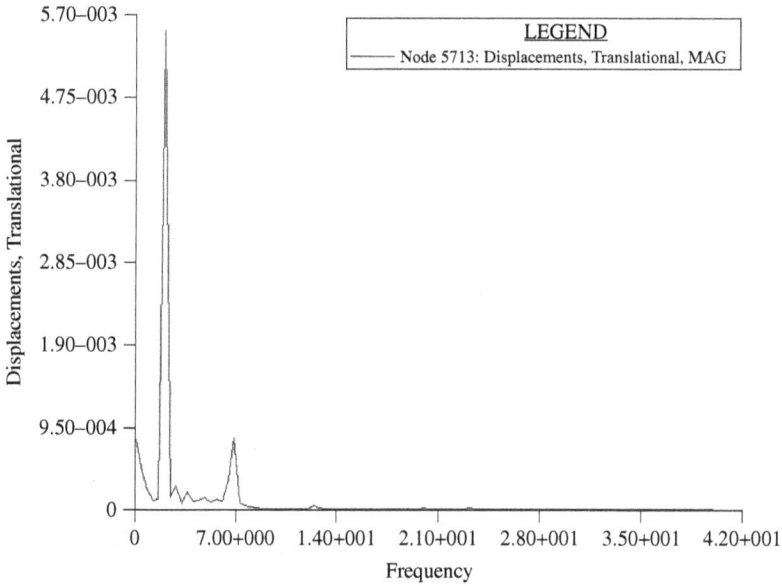

Figure 4.83. Displacement response spectrum at node F (enclosure).

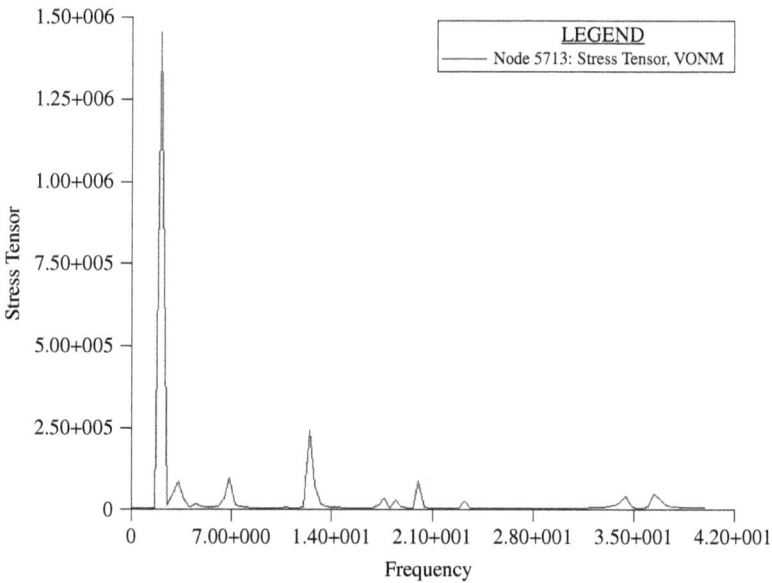

Figure 4.84. von Mises equivalent stress response spectrum at node F (enclosure).

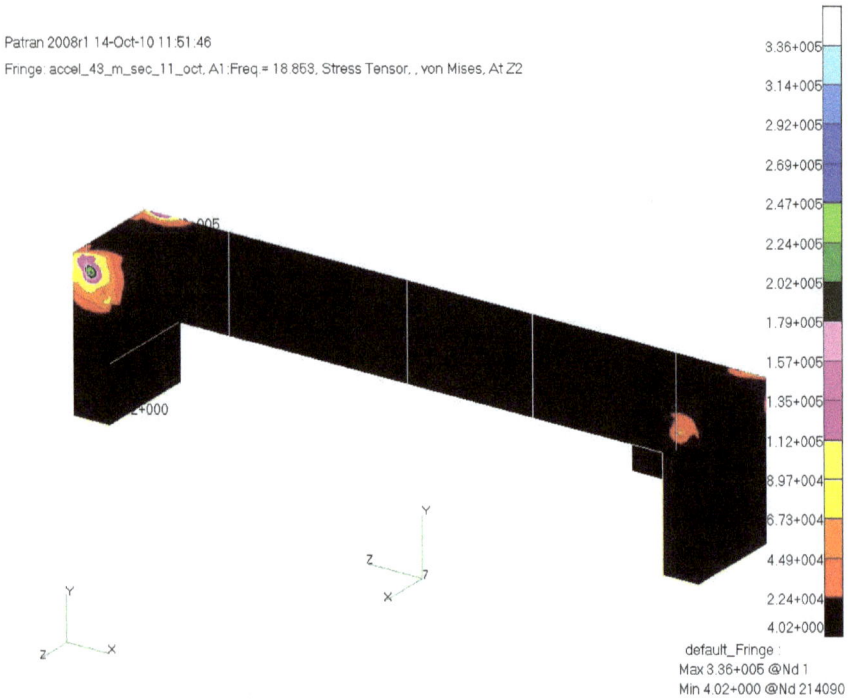

Figure 4.85. von Mises stresses in bus duct.

Table 4.20. Critical stress values and permissible limits.

Node position	Node no.	Max. stress (MPa)	Allowable stress (MPa)	% Stressed
K1	9	27	80	33.75
Insulator supporter	770	1.5	150	1
Insulator bar	780	4.8	—	
Enclosure	5713	1.5	80	1.8
Enclosure	8818	0.6	80	0.7
Channel	221718	24	80	30
Channel	234104	0.05	80	0.06
Channel	242806	0.09	80	0.11
Channel	273762	0.1	80	0.12

on the insulator, respectively. It is seen from the figures that the peak acceleration response of 1.25 m/s^2 occurs at 7 Hz where a near resonating-type displacement of amplitude 0.75 mm is seen. von Mises equivalent stress of about 4.8 MPa occurs closer to the natural frequency of the bus duct. Figures 4.78 and 4.79 show the acceleration, displacement and von Mises stress spectra at node E, which is the critical node on the channel. It is seen from the figures that the peak acceleration response of 1.5 m/s^2 occurs at 7 Hz and the near resonating-type displacement of amplitude 4.5 mm occurs at 2.5 Hz; von Mises equivalent stress of about 24 MPa occurs close to the natural frequency of the bus duct. Figure 4.80 shows the acceleration response spectrum at node F, which is the critical node on the enclosure. Figure 4.81 shows the displacement response and Figure 4.82 shows the von Mises stress response spectra at the same node. It is seen from the figures that the peak acceleration response of 1.5 m/s^2 occurs at 7 Hz while the displacement of amplitude 5.7 mm occurs at 3 Hz; von Mises equivalent stress of about 1.5 MPa occurs close to the natural frequency of the bus duct.

For the applied combination of input floor spectrum, the acceleration response peak occurs at a frequency closer to the natural frequency, indicating a near resonance type. Maximum displacement also occurs closer to the natural frequency at identified critical nodes. von Mises stress contours of the bus duct are shown in Figure 4.83. Figures 4.84 and 4.85 show von-Mises stress response at Node F and on the bus duct, respectively. It is seen from the figure that the critical stress occurs at node A. Maximum stress is limited to 0.34 N/mm^2, occurring at 18 Hz, which is lesser than that of the allowable stress limits shown in Table 4.20. It is seen from the analysis that the peak acceleration response, displacement amplitude and von Mises stresses are well within the safe permissible limits of the respective members. As seen in the table, critical stress values of various members are within the safe permissible limits.

References

Abdel-Jaber, H. and Glisic, B. 2016. Structural health monitoring methods for the evaluation of pre-stressing forces and pre-release cracks. *Frontiers in Built Environment*, *2*, p. 20.

Agdas, D., Rice, J.A., Martinez, J.R. and Lasa, I.R. 2015. Comparison of visual inspection and structural-health monitoring as bridge condition assessment methods. *Journal of Performance of Constructed Facilities*, *30*(3), p. 04015049.

Ahmed, M., Abdel, G. and Robert, H.S. 1985. Ambient vibration studies of golden gate bridge: I. Suspended structure. *Journal of Engineering Mechanics*, *111*(4), pp. 463–482.

Alippi, C., Anastasi, G., Di Francesco, M. and Roveri, M. 2010. An adaptive sampling algorithm for effective energy management in wireless sensor networks with energy-hungry sensors. *IEEE Transactions on Instrumentation and Measurement*, *59*(2), pp. 335–344.

Amafabia, D.M., Montalvão, D., David-West, O. and Haritos, G. 2018. A review of structural health monitoring techniques as applied to composite structures. *SDHM Structural Durability and Health Monitoring*, *11*(2), pp. 91–147.

Balageas, D., Fritzen, C.P. and Güemes, A. (eds.) 2010. *Structural Health Monitoring*, Vol. 90, John Wiley & Sons.

Begg, R., Mackenzie, A., Dodds, C. and Loland, O. 1976. Structural integrity monitoring using digital processing of vibration signals. *Proceedings of the 8th Annual Offshore Technology Conference*, pp. 305–311, 3–6 May, Houston, TX.

Benedetti, M., Fontanari, V. and Zonta, D. 2011. Structural health monitoring of wind towers: Remote damage detection using strain sensors. *Smart Materials and Structures*, *20*(5), p. 055009.

Bisht, S. 2005. *Methods for structural health monitoring and damage detection of civil and mechanical systems,* Ph.D. Dissertation, Virginia Tech.

Bozorgnia, Y. 2003. An introduction to the classic paper: A mechanical analyzer for the prediction of earthquake stresses, *Seismological Research Letters*, *74*, p. 312.

Bozorgnia, Y. and Campbell, K.W. 2004. The vertical-to-horizontal response spectral ratio and tentative procedures for developing simplified V/H and vertical design spectra, *Journal of Earthquake Engineering, 8*, pp. 175–207.

Bremer, K., Wollweber, M., Weigand, F., Rahlves, M., Kuhne, M., Helbig, R. and Roth, B. 2016. Fibre optic sensors for the structural health monitoring of building structures. *Procedia Technology, 26*, pp. 524–529.

Brincker, R., Kirkegaard, P. and Andersen, P. 1995. Damage detection in an offshore structure. *Fracture and Dynamics, 56*, R9434.

Brownjohn, J.M. 2007. Structural health monitoring of civil infrastructure. *Philosophical Transactions of the Royal Society of London A: Mathematical, Physical and Engineering Sciences 365*(1851), pp. 589–622.

Brownjohn, J.M., De Stefano, A., Xu, Y.L., Wenzel, H. and Aktan, A.E. 2011. Vibration-based monitoring of civil infrastructure: Challenges and successes. *Journal of Civil Structural Health Monitoring, 1*(3–4), pp. 79–95.

Caicedo, J.M., Dyke, S.J. and Johnson, E.A. 2004. Natural excitation technique and eigensystem realization algorithm for phase I of the IASC-ASCE benchmark problem: Simulated data. *Journal of Engineering Mechanics, 130*(1), pp. 49–60.

Campbell, K.W. and Bozorgnia, Y. 2003. Updated near-source ground motion (attenuation) relations for the horizontal and vertical components of peak ground acceleration and acceleration response spectra, *Bulletin of the Seismological Society of America, 93*, pp. 314–331.

Carden, E.P. and Fanning, P. 2004. Vibration based condition monitoring: A review. *Structural Health Monitoring, 3*(4), pp. 355–377.

Catbas, F.N., Zaurin, R., Gul, M. and Gokce, H.B. 2011. Sensor networks, computer imaging, and unit influence lines for structural health monitoring: Case study for bridge load rating. *Journal of Bridge Engineering, 17*(4), pp. 662–670.

Çelebi, M. 2002. *Seismic Instrumentation of Buildings, Special GSA. USGS Project: An Administrative Report*, United States Geological Survey, Menlo Park, CA.

Çelebi, M., Sanli, A., Sinclair, M., Gallant, S., and Radulescu, D. 2004. Real-time seismic monitoring needs of a building owner and the solution — A cooperative effort. *Journal of EERI, Earthquake Spectra, 19*(1), pp. 1–23.

Chandraskeran, S. 2013a. Advanced marine structures, Video course on NPTEL portal. Available at http://nptel.ac.in/courses/114106037/.

Chandraskeran, S. 2013b. Dynamics of ocean structures, Video course on NPTEL portal. Available at http://nptel.ac.in/courses/114106036/.

Chandraskeran, S. 2013c. Health, safety and environmental management (HSE) for oil and gas industries, Video course on NPTEL portal. Available at http:// nptel.ac.in/courses/114106017/.

Chandraskeran, S. 2013d. Ocean structures and materials, Video course on NPTEL portal. Available at http://nptel.ac.in/courses/114106035/.

Chandrasekaran, S. 2014. Advanced theory on offshore plant FEED engineering, Changwon National University, Republic of South Korea, p. 237.

Chandraskeran, S. 2015a. Dynamic analysis of offshore structures, Video course on MOOC, NPTEL portal. Available at https://onlinecourses.nptel.ac.in/ noc15_oe01/preview.

Chandraskeran, S. 2015b. HSE for offshore and petroleum engineering, Video course on MOOC, NPTEL portal. Available at https://onlinecourses.nptel.ac.in/noc15_oe02.

Chandrasekaran, S. 2015c. *Advanced Marine Structures*, CRC Press, FL.

Chandrasekaran. S. 2016a. *Health, Safety and Environmental Management for Offshore and Petroleum Engineers*, John Wiley and Sons, UK.

Chandrasekaran. S. 2016b. *Offshore Structural Engineering: Reliability and Risk Assessment*, CRC Press, FL.

Chandraskeran, S. 2016c. Risk and reliability of offshore structures, Video course on MOOC, NPTEL portal. Available at https://onlinecourses.nptel.ac.in/noc16_oe01.

Chandraskeran, S. 2016d. Safety practices for offshore and petroleum engineers, Video course on MOOC, NPTEL portal. Available at https://onlinecourses.nptel.ac.in/noc16_oe02.

Chandrasekaran, S. 2017a. *Dynamic Analysis and Design of Ocean Structures*, 2nd Edition, Springer, Singapore.

Chandraskeran, S. 2017b. Dynamics of ocean structures, Video course under MOOC, NPTEL portal, https://onlinecourses.nptel.ac.in/noc15_oe01/.

Chandraskeran, S. 2017c. Offshore structures under special loads including fire resistance, Video course under MOOC, NPTEL portal. Available at http://nptel.ac.in/courses/114106043.

Chandraskeran, S. 2018a. Computer methods of analysis of offshore structures, Video course under MOOC, NPTEL portal. Available at http://nptel.ac.in/courses/114106045.

Chandraskeran, S. 2018b. Safety practices for offshore and petroleum engineers, Video course re-run under MOOC.

Chandraskeran, S. 2018c. Structural health Monitoring, Video course under MOOC, NPTEL portal. Available at https://onlinecourses.nptel.ac.in/noc18_oe05/preview.

Chandrasekaran, S. and Bhattacharyya, S.K. 2012. Analysis and design of offshore structures with illustrated examples. *Human Resource Development Centre for Offshore and Plant Engineering (HOPE)*, Changwon National University Press.

Chandrasekaran, S. and Gaurav, S. 2017. *Design Aids for Offshore Structures under Special Environmental Loads Including Fire Resistance*, Springer, Singapore.

Chandrasekaran, S. and Jain, A.K. 2016. *Ocean Structures: Construction, Materials and Operations*, CRC Press, FL.

Chandrasekaran, S. and Lognath, R.S. 2017. Dynamic analyses of buoyant leg storage regasification platform (BLSRP) under regular waves: Experimental investigations, *Ships and Offshore Structures*, *12*(2), pp. 227–232.

Chandrasekaran, S. and Madhavi, N. 2015a. Retrofitting of offshore cylindrical structures with different geometrical configuration of perforated outer cover, *International Journal of Shipbuilding Progress*, *62*(1–2), pp. 43–56.

Chandrasekaran, S. and Madhavi, N. 2015b. Design aids for offshore structures with perforated members, *Ship and Offshore Structures*, *10*(2), pp. 183–203.

Chandrasekaran, S. and Merin, T. 2016. Suppression system for offshore cylinders under vortex induced vibration, *Vibro-engineering Procedia, 7,* pp. 01–06.

Chandrasekaran, S. and Roy, A. 2004. Comparison of modal combination rules in seismic analysis of multi-storey RC frames, *Proceedings of 3rd International Conference on Vibration Engineering & Technology of Machinery and 4th Asia Pacific Conference on System Integrity & Maintenance (Vetomac-3),* IIT-Kanpur, India, December 6–9, pp. 161–169.

Chandrasekaran, S. and Roy, A. 2006. Seismic evaluation of multi-storey RC frames using modal pushover analysis, *International Journal of Nonlinear Dynamics, 43*(4), pp. 329–342.

Chandrasekaran, S. and Tripati, U.K. 2004. Seismic vulnerability of irregular buildings, *Proceedings of 2nd International Conference on Protection of Structures Against Hazards,* Singapore, December 1–3, pp. 129–137.

Chandrasekaran, S. and Tripati, U.K. 2005. Geometric irregularity effects on seismic vulnerability of buildings, *Advances in Vibration Engineering, 4*(2), pp. 115–123.

Chandraskeran, S. and Thailammai, C. 2015. Structural health monitoring of offshore structures using wireless sensor networking under operational and environmental variability, *International Journal of Environmental, Chemical, Ecological, Geological and Geophysical Engineering, 10*(1), pp. 33–39.

Chandrasekaran, S. and Thailammai, C. 2016. Health monitoring of offshore structures using wireless sensor network: Experimental investigations. *Proceedings of SPIE 9804, Non-destructive Characterization and Monitoring of Advanced Materials, Aerospace, and Civil Infrastructure,* 980416, April 8.

Chandrasekaran, S. and Yuvraj, K. 2013. Dynamic analysis of a Tension Leg Platform under extreme waves, *Journal of Naval Architecture and Marine Engineering, 10,* p.5968.

Chandrasekaran, S., Anchuri, P.K. and Dubey, A. 2005. Seismic vulnerability of asymmetric reinforced concrete framed buildings, *Proceedings of International Convention of Structural Engineering Convention (SEC 2005),* IISc Bangalore.

Chandrasekaran, S., Dubey, A.K. and Tripati, U.K. 2006. Seismic behavior of SMRF under various structural irregularities, *Proceedings of International Conference on Earthquake Engineering,* School of Civil Engineering, Tanjavur, Tamilnadu, pp. 139–148.

Chandrasekaran, S., Ranjani, R. and Deepak, K. 2017. Response control of tension leg platform with passive damper: Experimental investigations, *Ships and Offshore Structures, 12*(2), pp. 171–181.

Chandrasekaran, S., Seriono, G. and Varun, G. 2007. Performance evaluation and damage assessment of buildings subjected to seismic loading, *Proceedings of 6th International Conference on Earthquake Resistant Engineering Structures (EERS-2007),* Bologna, Italy.

Chandrasekaran, S., Serino, G. and Varun, G. 2008. Performance evaluation assessment of buildings under seismic loading, *Proceedings of 10th International Conference on Structures Under Shock and Impact (SUSI-2008)* May 14–16, Portugal, pp. 313–322.

Chandrasekaran, S., Thailammai, C. and Shihas, A.K. 2016. Structural health monitoring of offshore structures using wireless sensor networking under operational and environmental variability, *International Journal of Environmental, Chemical and Ecological Engineering*, *10*(1), pp. 33–39.

Chandrasekaran, S., Nunziante, L., Serino, G. and Carannante, F. 2009. *Seismic Design Aids for Nonlinear analysis of Reinforced Concrete Structures*, CRC Press, FL.

Chandrasekaran, S., Nunziante, L., Serino, G. and Carannante, F. 2010. Axial force-Bending moment limit domain and Flow Rule for reinforced concrete elements using Euro Code *International Journal of Damage Mechanics*, *19*, pp. 523–558.

Chandrasekaran, S., Nunziante, L., Serino, G. and Carannante, F. 2011. Curvature ductility of RC sections based on Euro Code: Analytical procedure, *Journal of Civil Engineering*, Korean Society of Civil Engineers, Springer, *15*(1), pp. 131–144.

Chandrasekaran, S., Nunziante, L., Varun, G. and Carannante, F. 2008. Nonlinear seismic analyses of high rise reinforced concrete buildings, *ICFAI Journal of Structural Design*, *1*(1), pp. 7–24.

Chandrasekaran, S., Jain, A.K., Serino, G., Spizzuoco, M., Srivastava, S. and Varun, G. 2007. Risk assessment of seismic vulnerabilities of RC framed buildings due to asymmetricity, *Proceedings of 8th Pacific Conference on Earthquake Engineering*, Singapore.

Chen, Z., Zhou, X., Wang, X., Dong, L. and Qian, Y. 2017. Deployment of a smart structural health monitoring system for long-span arch bridges: A review and a case study. *Sensors*, *17*(9), p. 2151.

Chung, H.C., Enomoto, T., Shinozuka, M., Chou, P., Park, C., Yokoi, I. and Morishita, S. 2004. Real-time visualization of structural response with wireless MEMS sensors, *Proceedings of 13th World Conference on Earthquake Engineering*, Vancouver, BC, Canada, August, Vol. 121, pp. 1–10.

Ciang, C.C., Lee, J.R. and Bang, H.J. 2008. Structural health monitoring for a wind turbine system: A review of damage detection methods. *Measurement Science and Technology*, *19*(12), p. 122001.

Daniele, I. and Roberto, W. 2011. Integrated structural health monitoring system for high-rise buildings, *First Middle East Conference on Smart Monitoring, Assessment and Rehabilitation of Civil Structures*, Dubai, UAE, February, pp. 8–10.

Dastan Diznab, M.A., Mohajernassab, S., Seif, M.S., Tabeshpour, M.R. and Mehdigholi, H. 2014. Assessment of offshore structures under extreme wave conditions by modified endurance wave analysis. *Marine Structures*, *39*, pp. 50–69.

De Medeiros, R., Lopes, H.M., Guedes, R.M., Vaz, M.A., Vandepitte, D. and Tita, V. 2015. A new methodology for structural health monitoring applications. *Procedia Engineering*, *114*, pp. 54–61.

Di Lorenzo, E., Manzato, S., Peeters, B., Marulo, F. and Desmet, W. 2017. Structural health monitoring strategies based on the estimation of modal parameters. *Procedia Engineering*, *199*, pp. 3182–3187.

Ding, Y.L., Wang, G.X., Sun, P., Wu, L.Y. and Yue, Q. 2015. Long-term structural health monitoring system for a high-speed railway bridge structure. *The Scientific World Journal, 2015*, p. 250562.

Do, R. 2014. *Passive and Active Sensing Technologies for Structural Health Monitoring*. University of California, San Diego.

Doebling, S.W., Farrar, C.R. and Prime, M.B. 1998. A summary review of vibration-based damage identification methods. *Shock and Vibration Digest, 30*(2), pp. 91–105.

Doebling, S.W., Farrar, C.R., Prime, M.B. and Shevitz, D.W. 1996. *Damage Identification and Health Monitoring of Structural and Mechanical Systems from Changes in Their Vibration Characteristics: A Literature Review*. No. LA-13070-MS. Los Alamos National Lab, NM.

Dubbs, N.C. and Yarnold, M. 2014. Optimal sensor placement for condition assessment of a cantilever truss bridge. *NDE/NDT for Structural Materials Technology for Highway & Bridges*, pp. 106–113.

Elshafey, A.A., Haddara, M.R. and Marzouk, H. 2009. Dynamic response of offshore jacket structures under random loads. *Marine Structures, 22*(3), pp. 504–521.

Fan, W. and Qiao, P. 2011. Vibration-based damage identification methods: A review and comparative study. *Structural Health Monitoring, 10*(1), pp. 83–111.

Farahani, R.V. and Penumadu, D. 2016. Damage identification of a full-scale five-girder bridge using time-series analysis of vibration data. *Engineering Structures, 115*, pp. 129–139.

Farrar, C.R. and Worden, K. 2007. An introduction to structural health monitoring. *Philosophical Transactions of the Royal Society of London A: Mathematical, Physical and Engineering Sciences, 365*(1851), pp. 303–315.

Farrar, C.R., Allen, D.W., Park, G., Ball, S. and Masquelier, M.P. 2006. Coupling sensing hardware with data interrogation software for structural health monitoring. *Shock and Vibration, 13*(4–5), pp. 519–530.

Farrar, C.R., Baker, W.E., Bell, T.M., Cone, K.M., Darling, T.W., Duffey, T.A., Eklund, A. and Migliori, A. 1994. *Dynamic Characterization and Damage Detection in the I-40 Bridge Over the Rio Grande*. Report No. LA-12767-MS, Los Alamos National Laboratory, NM.

Federici, F., Alesii, R., Colarieti, A., Faccio, M., Graziosi, F., Gattulli, V. and Potenza, F. 2014. Design of wireless sensor nodes for structural health monitoring applications. *Procedia Engineering, 87*, pp. 1298–1301.

Flynn, E. 2010. *A bayesian experimental design approach to structural health monitoring with application to ultrasonic guided waves*. Ph.D. Dissertation, University of California, San Diego.

Gattulli, V. 2013. Advanced applications in the field of structural control and health monitoring after the 2009 L'Aquila earthquake. *Engineering Seismology, Geotechnical and Structural Earthquake Engineering*, InTech.

Geng, B.L., Teng, B. and Ning, D.Z. 2010. A time-domain analysis of wave force on small-scale cylinders of offshore structures. *Journal of Marine Science and Technology, 18*(6), pp. 875–882.

Giurgiutiu, V. and Cuc, A. 2005. Embedded non-destructive evaluation for structural health monitoring, damage detection, and failure prevention. *Shock and Vibration Digest*, *37*(2), p. 83.

Glisic, B. and Inaudi, D. 2008. *Fibre Optic Methods for Structural Health Monitoring*. John Wiley & Sons.

Hamburger, R. 2000. *A Policy Guide to Steel Moment-frame Construction*, Washington DC, Federal Emergency Management Agency, Technical Report No. 354.

Heo, G. and Jeon, J. 2009. A smart monitoring system based on ubiquitous computing technique for infra-structural system: Centering on identification of dynamic characteristics of self-anchored suspension bridge. *KSCE Journal of Civil Engineering*, *13*(5), pp. 333–337.

Hudson, D.E. 1979. Reading and interpreting strong motion accelerograms, *Engineering Monographs on Earthquake Criteria, Structural Design, and Strong Motion Records 1*, EERI, Berkeley, California.

Hudson, D.E. 1990. Reading and interpreting strong motion accelerograms *Engineering Monographs on Earthquake Criteria, Structural Design, and Strong Motion Records 1*. EERI.

Ibrahim, M.E. 2016. Non-destructive testing and structural health monitoring of marine composite structures. In *Marine Applications of Advanced Fibre-Reinforced Composites*, pp. 147–183.

Inaudi, D. 2010. Long-term static structural health monitoring. In *Structures Congress, 2010*, pp. 566–577.

Jahangiri, V., Mirab, H., Fathi, R. and Ettefagh, M.M. 2016. TLP structural health monitoring based on vibration signal of energy harvesting system. *Latin American Journal of Solids and Structures*, *13*(5), pp. 897–915.

Joao LDC 2003. Standard methods for seismic analyses, BYG.DTU R-064.

Keith, W., Mason, G. and Allman, D. 2003. Experimental validation of structural monitoring methodology: Part I Novelty detection on a laboratory structures. *Journal of Sound and Vibration*, *259*(2), pp. 323–343.

Kesavan, A., John, S. and Herszberg, I. 2008. Structural health monitoring of composite structures using artificial intelligence protocols. *Journal of Intelligent Material Systems and Structures*, *19*(1), pp. 63–72.

Khan, A.A., Zafar, S., Khan, N.S. and Mehmood, Z. 2014. History, current status and challenges to structural health monitoring system aviation field. *Space Technology*, *4*, pp. 67–74.

Kianian, M., Golafshani, A. and Ghodrati, E. 2013. Damage detection of offshore jacket structures using frequency domain selective measurements. *Journal of Marine Science Applications*, *13*, pp. 193–199.

Kim, S., Pakzad, S., Culler, D., Demmel, J., Fenves, G., Glaser, S. and Turon, M. 2007. Health monitoring of civil infrastructures using wireless sensor networks. *Proceedings of the 6th International Conference on Information Processing in Sensor Networks*, pp. 254–263.

Komachi, Y., Tabeshpour, M.R., Golafshani, A.A. and Mualla, I. 2011. Retrofit of Ressalat jacket platform (Persian Gulf) using friction damper device. *Journal of Zhejiang University-Science A*, *12*(9), pp. 68–691.

Kurata, M., Li, X., Fujita, K. and Yamaguchi, M. 2013. Piezoelectric dynamic strain monitoring for detecting local seismic damage in steel buildings. *Smart Materials and Structures*, *22*(11), p. 115002.

Lee, Y., Blaauw, D. and Sylvester, D. 2016. Ultralow power circuit design for wireless sensor nodes for structural health monitoring. *Proceedings of the IEEE*, *104*(8), pp. 1529–1546.

Li, B., Sun, Z., Mechitov, K., Hackmann, G., Lu, C., Dyke, S.J., Agha, G. and Spencer, B.F. 2013, April. Realistic case studies of wireless structural control. *2013 ACM/IEEE International Conference on IEEE: Cyber-Physical Systems (ICCPS)*, pp. 179–188.

Li, H., Wang, J. and Hu, S.L.J. 2008. Using incomplete modal data for damage detection in offshore jacket structures. *Ocean Engineering*, *35*(17), pp. 1793–1799.

Li, H.N., Ren, L., Jia, Z.G., Yi, T.H. and Li, D.S. 2016. State-of-the-art in structural health monitoring of large and complex civil infrastructures. *Journal of Civil Structural Health Monitoring*, *6*(1), pp. 3–16.

Liu, W.Y., Tang, B.P., Han, J.G., Lu, X.N., Hu, N.N. and He, Z.Z. 2015. The structure healthy condition monitoring and fault diagnosis methods in wind turbines: A review. *Renewable and Sustainable Energy Reviews*, *44*, pp. 466–472.

Liu, Y. and Nayak, S. 2012. Structural health monitoring: State of the art and perspectives. *Journal of the Minerals, Metals & Materials Society*, *64*(7), pp. 789–792.

Loland, O. and Dodds, C.J. 1976. Experiences in developing and operating integrity monitoring systems in the North Sea. *Proceedings of the 8th Annual Offshore Technology Conference*, Houston, TX, May 3–6, 1976, No. V2, pp. 313–320.

Lynch, J.P. and Loh, K.J. 2006. A summary review of wireless sensors and sensor networks for structural health monitoring. *Shock and Vibration Digest*, *38*(2), pp. 91–130.

Lynch, J.P., Law, K.H., Kiremidjian, A.S., Kenny, T.W., Carryer, E. and Partridge, A. 2001. The design of a wireless sensing unit for structural health monitoring. *Proceedings of the 3rd International Workshop on Structural Health Monitoring*, Stanford, CA, September 12–14.

Lynch, J.P., Law, K.H., Straser, E.G., Kiremidjian, A.S. and Kenny, T.W. 2000. The development of a wireless modular health monitoring system for civil structures. *MCEER Mitigation of Earthquake Disaster by Advanced Technologies (MEDAT-2) Workshop*.

Lynch, J.P., Sundararajan, A., Law, K.H., Kiremidjian, A.S. and Carryer, E. 2004. Embedding damage detection algorithms in a wireless sensing unit for operational power efficiency. *Smart Materials and Structures*, *13*(4), p. 800.

Lynch, J.P., Swartz, R.A., Zimmerman, A.T., Brady, T.F., Rosario, J., Salvino, L.W. and Law, K.H. 2009. Monitoring of a high speed naval vessel using a wireless hull monitoring system. *Proceedings of the 7th*

International Workshop on Structural Health Monitoring, September 2009, pp. 9–11.

Márquez, F.P.G., Tobias, A.M., Pérez, J.M.P. and Papaelias, M. 2012. Condition monitoring of wind turbines: Techniques and methods. *Renewable Energy*, *46*, pp. 169–178.

Martinez-Luengo, M., Kolios, A. and Wang, L. 2016. Structural health monitoring of offshore wind turbines: A review through the statistical pattern recognition paradigm. *Renewable and Sustainable Energy Reviews*, *64*, pp. 91–105.

Mieloszyk, M. and Ostachowicz, W. 2017. An application of structural health monitoring system based on FBG sensors to offshore wind turbine support structure model. *Marine Structures*, *51*, pp. 65–86.

Mohammadi, M.E., Yousefianmoghadam, S., Wood, R.L. and Stavridis, A. 2016. Damage quantification from point clouds for finite element model calibration. *SEI/ASCE Structural Health Monitoring Applications Case Study Archive*.

Mollineaux, M., Balafas, K., Branner, K., Nielsen, P., Tesauro, A., Kiremidjian, A. and Rajagopal, R. 2014. Damage detection methods on wind turbine blade testing with wired and wireless accelerometer sensors. *EWSHM-7th European Workshop on Structural Health Monitoring*.

Moss, R.M. and Matthews, S.L. 1995. In-service structural monitoring: A state of the art review. *Structural Engineer*, *73*(2), pp. 23–31.

Mourad, S.A., Sadek, A.W. and Batisha, A.F. 2001. Structural health monitoring of offshore structures. *Proceedings of the 6th International Conference on the Application of Artificial Intelligence to Civil and Structural Engineering*, Civil-Comp Press, p. 39.

Nagayama, T. and Spencer, Jr. B.F. 2007. *Structural Health Monitoring Using Smart Sensors*. Newmark Structural Engineering Laboratory, University of Illinois at Urbana-Champaign.

Neves, A.C., González, I., Leander, J. and Karoumi, R. 2017. Structural health monitoring of bridges: A model-free ANN-based approach to damage detection. *Journal of Civil Structural Health Monitoring*, *7*(5), pp. 689–702.

Nguyen, V.H., Schommer, S., Zürbes, A. and Maas, S. 2016. Structural health monitoring based on static measurements with temperature compensation. *Quality Specifications for Roadway Bridges, Standardization at a European Level*.

Noman, A.S., Deeba, F. and Bagchi, A. 2009. Structural health monitoring using vibration-based methods and statistical pattern recognition techniques. *CANSMART 2009 International Workshop Smart Materials and Structures*, pp. 22–23.

Okoh, P. and Haugen, S. 2013. The influence of maintenance on some selected major. *Chemical Engineering Transactions*, *31*, pp. 493–498.

Park, J.W., Jung, H.J., Jo, H. and Spencer, B.F. 2012. Feasibility study of micro-wind turbines for powering wireless sensors on a cable-stayed bridge. *Energies*, *5*(9), pp. 3450–3464.

Park, M.S., Koo, W. and Kawano, K. 2011. Dynamic response analysis of an offshore platform due to seismic motions. *Engineering Structures, 33*(5), pp. 1607–1616.

Peng, J., Hongbo, X., Zhiye, H. and Wang, Z. 2009. Design of a water environment monitoring system based on wireless sensor networks. *Sensors, 9*(8), pp. 6411–6434.

Pérez, C.A., Jimenez, M., Soto, F., Torres, R., López, J.A. and Iborra, A. 2011. A system for monitoring marine environments based on wireless sensor networks. *OCEANS*, IEEE,-Spain, June 2011, pp. 1–6. IEEE.

Perišić, N., Kirkegaard, P.H. and Tygesen, U.T. 2014. Load identification of offshore platform for fatigue life estimation. *Structural Health Monitoring*, Vol. 5, Springer, Cham, pp. 99–109.

Pook, L. 1983. *The Role of Crack Growth in Metal Fatigue*. The Metal Society, London.

Reitherman, R. 1997. *Historic Developments in the Evolution of Earthquake Engineering*. CUREE.

Rice, J.A. and Spencer, B. 2009. *Flexible Smart Sensor Framework for Autonomous Full-scale Structural Health Monitoring*, NSEL Report Series, Report No. NSEL-018, Urbana.

Rice, J.A. and Spencer, Jr. B.F. 2008. Structural health monitoring sensor development for the Imote2 platform. *The 15th International Symposium on: Smart Structures and Materials & Non-destructive Evaluation and Health Monitoring*, International Society for Optics and Photonics.

Roghaei, M. and Zabihollah, A. 2014. An efficient and reliable structural health monitoring system for buildings after earthquake. *APCBEE Proceedia, 9*, pp. 309–316.

Rytter, A. 1993. *Vibration based inspection of civil engineering structures.* Ph.D. Dissertation, Department of Building Technology and Structural Engineering, University of Aalborg, Aalborg.

Shaladi, R., Alatshan, F. and Yang, C. 2015. An overview on the applications of structural health monitoring using wireless sensor networks in bridge engineering. *Proceedings of the International Conference Advanced Science and Engineering Technology on Natural Resources*, pp. 4–11.

Shrive, N.G. 2005. Intelligent structural health monitoring: A civil engineering perspective. *2005 IEEE International Conference on Systems, Man and Cybernetics*, Vol. 2, IEEE, pp. 1973–1977.

Smarsly, K. and Hartmann, D. 2007. Artificial intelligence in structural health monitoring. *Proceedings of the 3rd International Conference on Structural Engineering, Mechanics and Computation*, Cape Town, South Africa, Vol. 10, No. 09.

Smarsly, K., Dragos, K. and Wiggenbrock, J. 2016. Machine learning techniques for structural health monitoring. *Proceedings of the 8th European Workshop on Structural Health Monitoring (EWSHM 2016)*, Bilbao, Spain, pp. 5–8.

Sohn, H., Farrar, C.R., Hemez, F.M. and Czarnecki, J.J. 2002. *A Review of Structural Health Review of Structural Health Monitoring Literature 1996–2001*. No. LA-UR-02-2095. Los Alamos National Laboratory.

Spencer, B.F., Jo, H., Mechitov, K.A., Li, J., Sim, S.H., Kim, R.E., Cho, S., Linderman, L.E., Moinzadeh, P., Giles, R.K. and Agha, G. 2016. Recent advances in wireless smart sensors for multi-scale monitoring and control of civil infrastructure. *Journal of Civil Structural Health Monitoring*, *6*(1), pp. 17–41.

Spignesi, S.J. 2005. *Catastrophe! 100 Greatest Disasters of All Time*. Kensington Publishing Corp.Statista. 2015. Statista. retrieved from Statista: https://www.statista.com/statistics/279100/number-of-offshore-rigs-world wide-by-region.

Straser, E.G. and Kiremidjian, A.S. 1998. *A Modular, Wireless Damage Monitoring System for Structures*, Technical Report 128, John A. Blume Earthquake Engineering Center, Stanford University, Stanford, CA.

Sun, M., Staszewski, W.J. and Swamy, R.N. 2010. Smart sensing technologies for structural health monitoring of civil engineering structures. *Advances in Civil Engineering*, Article ID 724962, p. 13.

Taha, M.R. and Lucero, J. 2005. Damage identification for structural health monitoring using fuzzy pattern recognition. *Engineering Structures*, *27*(12), pp. 1774–1783.

Vamvatsikos, D. and Cornell, C.A. 2002. Incremental dynamic analysis, *Earthquake Engineering and Structural* Dynamics, *31*(3), pp. 491–514.

Vanik, M.W., Beck, J.L. and Au, S. 2000. Bayesian probabilistic approach to structural health monitoring. *Journal of Engineering Mechanics*, *126*(7), pp. 738–745.

Venanzi, I. 2016. A review on adaptive methods for structural control. *The Open Civil Engineering Journal*, *10*(1), pp. 12–24.

Wang, Y., Loh, K.J., Lynch, J.P., Fraser, M., Law, K.H. and Elgamal, A. 2006. Vibration monitoring of the Voigt Bridge using wired and wireless monitoring systems. *Proceedings of the 4th China–Japan–US Symposium on Structural Control and Monitoring*, pp. 16–17.

Wang, Y., Lynch, J. and Law, K. 2005. Validation of an integrated network system for real-time wireless monitoring of civil structures, *Proceedings of the 5th International Workshop on Structural Vibrations*, Stanford, CA.

Wilson, E. and Clough, R. 1999. Early finite element research at Berkeley, *Proc. of 5th National Conf in Computational Mech.*, TX, USA.

Wong, K. 2003. Structural identification of Tsing Ma Bridge. *Transactions Hong Kong Institution of Engineers*, *10*(1), pp. 38–47 .

Wymore, M.L., Van Dam, J.E., Ceylan, H. and Qiao, D. 2015. A survey of health monitoring systems for wind turbines. *Renewable and Sustainable Energy Reviews*, *52*, pp. 976–990.

Xiang, P., Nishitani, A. and Wu, M. 2017. Seismic vibration and damage control of high-rise structures with the implementation of a pendulum-type nontraditional tuned mass damper. *Structural Control and Health Monitoring*, *24*(12).

Yan, Y.J., Cheng, L., Wu, Z.Y. and Yam, L.H. 2007. Development in vibration-based structural damage detection technique. *Mechanical Systems and Signal Processing*, *21*(5), pp. 2198–2211.

Yarnold, M. and Dubbs, N.C. 2015. Bearing Assessment with periodic temperature-based measurements. *Journal of the Transportation Research Record, 2481,* pp. 115–123.

Yu, H., Ha, M.K., Choi, J.W. and Tai, J.S.C. 2006. Design and implementation of a comprehensive full-scale measurement system for a large container carrier. *Proceedings of the International Conference on Design and Operation of Container Ships (RINA),* London.

Yu, T., Twumasi, J.O., Le, V., Nonis, C., Reagan, D. and Niezrecki, C. 2016. Structural health monitoring of bridge abutments using imaging radar and digital image correlation. *Collection of SHM Case Studies by ASCE SEI Methods of Monitoring Committee Report.*

Yu, Y. and Ou, J. 2008. Wireless sensing experiments for structural vibration monitoring of offshore platform. *Frontiers of Electrical and Electronic Engineering in China, 3*(3), pp. 333–337.

Tutorials, Keys
and Test Papers

Tutorial Paper 1

Answer all the questions (Total marks: 20)

PART A (5 × 1 = 5)

1. _____ is required to understand the response behaviour of the offshore structures under loading conditions.
2. _____ deals with the assessment of actual conditions and load carrying capacity of the structural systems.
3. _____ is required to process the strain variation caused by the damages in local scale detected on the sensor coatings.
4. In imaging ultrasonic sensors, _____ generates signals that pass through the material.
5. If the system functionality is lost, it is called _____.

PART B (5 × 3 = 15)

1. Why is structural health monitoring (SHM) vital for offshore structures?
2. How can SHM be useful in detecting early risks?
3. What are the commonly used tools of SHM for the aviation industry?
4. Differentiate active and passive SHM.
5. Classify damages that can occur to a system with respect to the scale of damage.

Tutorial Paper 2

Answer all the questions (Total marks: 20)

PART A (5 × 1 = 5)

1. _____ is a vital component integrally connected to SHM.
2. Data acquisition depends on _____, _____ and _____.
3. _____ tracks the thermal load path in a material, travelled longitudinally over a point of time.
4. SHM system is a generic system. State True or False.
   If False, rewrite the correct statement.
5. _____ is the simpler way to address the uncertainties that arise in the SHM.

PART B (5 × 3 = 15)

1. List preliminary disadvantages of unsatisfactory maintenance of a structural system.
2. What are the levels of damage detection?
3. What are the different ways through which a statistical model can be developed?
4. Write short notes on the DIC technique.
5. State critical issues of uncertainties in SHM.

Tutorial Paper 3

Answer all the questions (Total marks: 20)

PART A (5 × 1 = 5)

1. _____ is the change in health conditions of a structural system in terms of its deterioration in performance (functional).
2. Sampling interval in triggered monitoring depends on _____ of the examined phenomenon.
3. _____ monitoring is carried out with a higher sampling rate.
4. Damage identification can be done through _____ redistribution.
5. In vibration-based damage detection, modal parameters are estimated on the basis of _____.

PART B (5 × 3 = 15)

1. Define long-term monitoring and periodic monitoring.
2. What are specific conditions under which long-term monitoring is preferably carried out?
3. Discuss the hypothesis of vibration-based damage detection.
4. List the global methods deployed in damage detection.
5. List important factors that influence maintenance of infrastructure, in general.

Tutorial Paper 4

Answer all the questions (Total marks: 20)

PART A (5 × 1 = 5)

1. _____ provides the base line for comparing response of a structure before and after damage.
2. Significant changes in the vibration characteristics can lead to _____.
3. Expression for damage sensitivity is not applicable for structures with _____.
4. Both the crack location and change in size depend on the changes in _____ and _____ of the structure before and after the damage.
5. Significant changes in _____ can enable a convenient damage detection process.

PART B (5 × 3 = 15)

1. What is the basic feature of vibration-based SHM?
2. List the steps involved in vibration-based SHM.
3. Draw the flow chart describing the vibration-based monitoring.
4. Define modal sensitivity.
5. What is the basic assumption in the method used to identify damage by element modal stiffness?

Tutorial Paper 5

Answer all the questions (Total marks: 20)

PART A ($5 \times 1 = 5$)

1. Visual inspection methods affect _____ and _____, significantly.
2. _____ is the sensor used for wind load assessment.
3. _____ methods are less contaminated by noise data.
4. Fractional change in the modal energy is related to _____, which occurs due to _____.
5. Curvature of mode shape can be approximated using _____ technique.

PART B ($5 \times 3 = 15$)

1. The results of visual inspection (VI) may be inadequate to compare with true assessment. Why?
2. List the factors that affect system complexity in SHM.
3. How can the maintenance of the SHM system be reduced?
4. What are the limitations of SCCM?
5. What are different stages involved in the long-term SHM?

Tutorial Paper 6

Answer all the questions (Total marks: 20)

PART A (5 × 1 = 5)

1. _____ are guided waves that travel in layered material.
2. PWAS is an active sensor. State True or False.
3. In flat plates, ultrasonic guided waves travel as _____ and _____.
4. In the pitch-catch method, flaws are detected by _____ and _____ due to damage.
5. Cracks are generally initiated in composites by _____ and _____.

PART B (5 × 3 = 15)

1. List the drawbacks (if any) of deploying ultrasonic method.
2. Explain passive SHM.
3. List the advantages of PWAS.
4. Discuss the applications of embedded NDE using the pitch-catch method.
5. Express the relationship used to determine probability of crack detection by the pitch-catch method.

Tutorial Paper 7

Answer all the questions (Total marks: 20)

PART A (5 × 1 = 5)

1. In the very early stage of monitoring, _____ sensors are used.
2. The period of monitoring during reconstruction is _____.
3. Fibre optic sensors use _____ to read and measure data.
4. Magnetostrictive sensors emit ultrasonic energy with very high strength. State True or False.
5. MEM sensors are manufactured using _____ technology.

PART B (5 × 3 = 15)

1. List the structures which require monitoring during the construction stage.
2. How can the FOS be used in the measurement of moisture ingression?
3. What is the magnetostrictive effect?
4. Mention the advantages of smart sensors.
5. Why is SHM mandatory for offshore structures?

Tutorial Paper 8

Answer all the questions (Total marks: 20)

PART A (5 × 1 = 5)

1. In areas where visual inspection is not possible, the members can be examined through _____.
2. The damage can be quantified before developing the sensing system by _____.
3. Wireless sensors convert the measured data into _____ for transmission.
4. In wired sensor networking, the central server is responsible for _____, _____, and _____ of data.
5. What is the most important advantage of wireless sensor network?

PART B (5 × 3 = 15)

1. List the factors to be considered while designing the monitoring system for offshore platforms.
2. What are the common factors that govern the selection of sensors?
3. List the advantages of implementing SHM in strategic structures.
4. What are the limitations of wired sensors?
5. List the advantages of MEMS sensor.

Tutorial Paper 9

Answer all the questions (Total marks: 20)

PART A (5 × 1 = 5)

1. One of the major challenges in the SHM design of offshore platforms is the _____ of the sensors.
2. BLSRP is _____ in the vertical plane and _____ in the horizontal plane.
3. What is the primary objective of the SHM design?
4. _____ are used to assess the errors in the collected data.
5. ANN combined with pre-processing tools, such as _____, can be used for damage detection.

PART B (5 × 3 = 15)

1. Why is VI not possible in offshore structures?
2. What are the components of the SHM system?
3. List some of the issues related to vibration-based damage detection.
4. List the design parameters to be considered in the wireless communication channels.
5. What are the advantages of GFRP?

Tutorial Paper 10

Answer all the questions (Total marks: 20)

PART A (5 × 1 = 5)

1. What is the main aim of embedded sensors?
2. _____ can enable the pattern for future prediction.
3. Unsupervised learning is very appropriate to handle SHM of complex system. State True or False.
4. What is the primary objective in the conceptual design of the SHM system?
5. _____ is used to process the acquired signal from the accelerometer module.

PART B (5 × 3 = 15)

1. What are the salient features of machine learning?
2. What is data mining?
3. List the factors that affect the output of the SHM system.
4. What are the issues that may arise in using wireless sensors over a large area?
5. What are the sources of noise in signal?

Tutorial Paper 11

Answer all the questions (Total marks: 20)

PART A (5 × 1 = 5)

1. In wireless sensor networking, power consumption can be reduced by decreasing the _____.
2. _____ can enhance the communication efficiency because they can handle _____.
3. _____ generates the reliable report when the acquired data exceed the present threshold value.
4. The tethers in tension Leg Platforms are slack moored. State True or False.
5. _____ power mode is used to reduce the power consumption in the SHM processing unit.

PART B (5 × 3 = 15)

1. How is the sampling rate chosen in SHM?
2. What are the different power modes in the processing unit of SHM?
3. What are the alternate sources of power supply for wireless sensors in real-time monitoring?
4. What are the advantages of using SHM designed with IEEE 802.11 protocol at an operating frequency of 2.4 GHz?
5. Explain the WSN architecture with a neat sketch.

Tutorial Paper 12

Answer all the questions (Total marks: 20)

PART A ($5 \times 1 = 5$)

1. Frequency-domain technique is used to analyse *stationary event*, which is localised in _____.
2. _____ is one of the best tools to identify the frequency component present in the signal.
3. What is the major defect in FFT?
4. _____ is the multivariate statistical model that can reduce the effect of environmental factors on the sensor performance.
5. Distributed health monitoring system is used in _____.

PART B ($5 \times 3 = 15$)

1. How are postulated failure cases introduced in TLP experimental study?
2. Explain the feature extraction process.
3. What are the tools used to analyse the data in frequency domain?
4. What are the drawbacks of VI?
5. What are the recent advancements in SHM technologies?

Tutorial Paper 1: Key

Answer all the questions (Total marks: 20)

PART A ($5 \times 1 = 5$)

1. *Continuous monitoring* is required to understand the response behaviour of the offshore structures under loading conditions.
2. *Structural assessment* deals with the assessment of actual conditions and load carrying capacity of the structural systems.
3. *A detailed spectroscopic analysis* is required to process the strain variation caused by the damages in local scale detected on the sensor coatings.
4. In imaging ultrasonic sensors, *ultrasonic wave transducer* generates signals that pass through the material.
5. If the system functionality is lost, it is called *failure*.

PART B ($5 \times 3 = 15$)

1. Why is structural health monitoring (SHM) vital for offshore structures?

   - *Maintenance is vital for self-operating unmanned platforms.*
   - *Structural repair should be carried out without shutting down the system.*
   - *Offshore structures need to be repaired when they are loaded.*

2. How can SHM be useful in detecting early risks?

   - *SHM can be deployed to detect a power structure and therefore its usage can be limited, ensuring public safety.*
   - *SHM is a useful tool in preventing water and flood damage caused by failure of dams and large reservoirs.*
   - *Built-in sensors are useful to monitor change in the water level and detect minor leaks in major failure.*

3. What are the commonly used tools of SHM for the aviation industry?

   - *Fuzzy pattern recognition;*
   - *Neural networks;*
   - *Diffused ultrasonic waves technique to detect the structural damage;*
   - *Vibration-based technique;*
   - *Intelligent parameter ranging technique for location of damage.*

4. Differentiate active and passive SHM.

   Passive SHM: When observing a structure as it evolves, a physical parameter and its state evolve as a result of interaction with the environment. Example: Acoustic emission.
   Active SHM: Structure is equipped both with sensors and actuators which prompt forces opposite to the structure. Example: Boeing 787 Dreamliner equipped with embedded sensors for continuous health monitoring.

5. Classify damages that can occur to a system with respect to the scale of damage.

   - *Long-term scale: corrosion, fatigue;*
   - *Short-term scale: due to impact load or shock load and aircraft landing.*

Tutorial Paper 2: Key

Answer all the questions (Total marks: 20)

PART A (5 × 1 = 5)

1. *Non-destructive evaluation* is a vital component integrally connected to SHM.
2. Data acquisition depends on *excitation methods, data transmission and sensing the structural response.*
3. *Infrared imaging* tracks the thermal load path in a material, travelled longitudinally over a point of time.
4. SHM system is a generic system. State True or False. *False.*
   It is problem-specific, accounting for certain classes of uncertainties related to the chosen system.
5. *Monte Carlo technique* is the simpler way to address the uncertainties that arise in the SHM.

PART B (5 × 3 = 15)

1. List preliminary disadvantages of unsatisfactory maintenance of a structural system.

   - *It may cause further disaster. Example: accident of Aloha Airlines.*
   - *Efficient use of funds towards maintenance is reduced.*
   - *Schedule of maintenance period can result in down-time of the facility at its critical need.*

2. What are the levels of damage detection?

- *Level 1: Determination of damage on the structure;*
- *Level 2: Determination of geometric location of the damage;*
- *Level 3: Quantification of severity of damage;*
- *Level 4: Prediction of remaining service life of the structure.*

3. What are the different ways through which a statistical model can be developed?

- *Learning under supervision: response surface analysis, neural networks, genetic algorithm;*
- *Learning under unmanned condition: control chart analysis, neural network, hypothesis testing.*

4. Write short notes on the DIC technique.

- *It is useful in detecting microcracking in the chopped fibre glass compressive moulded parts.*
- *It shows principal strains in the damaged regions where cracks are formed.*
- *It is useful in detecting localised residual stresses which are caused in the material upon removal of the load.*
- *It is used to track the strain variations that occur under temperature variations.*

5. State critical issues of uncertainties in SHM.

- *It can arise from parametric data, which arise from physical experiment and numerical simulation output.*
- *Imperfect knowledge of control parameters of the physical experiment and numerical simulation.*
- *It can also arise from stochastic equations of motion, environmental variations, measurement errors and numerical errors.*

Tutorial Paper 3: Key

Answer all the questions (Total marks: 20)

PART A ($5 \times 1 = 5$)

1. *Damage* is the change in health conditions of a structural system in terms of its deterioration in performance (functional).
2. Sampling interval in triggered monitoring depends on the *dynamic nature* of the examined phenomenon.
3. *Dynamic* monitoring is carried out with a higher sampling rate.
4. Damage identification can be done through *dead load* redistribution.
5. In vibration-based damage detection, the modal parameters are estimated on the basis of *vibration data*.

PART B ($5 \times 3 = 15$)

1. Define long-term monitoring and periodic monitoring.

   Long-term monitoring: It is a process of periodic or continuous monitoring, which is carried out over several years.
   Periodic monitoring: It is non-continuous monitoring, which is carried out to identify any significant change or detrimental damage on the structural system.

2. What are the specific conditions under which long-term monitoring is preferably carried out?

   - *If the changes in loading are slow, such as gradual change in temperature;*
   - *To predict the effect of natural hazards on the structural systems.*

3. Discuss the hypothesis of vibration-based damage detection.

   Structural damage can be characterised by local modification of stiffness, which in turn affects the modal parameters.

4. List the global methods deployed in damage detection.

   - *Natural frequency method;*
   - *Mode shape and operational frequency method;*
   - *Modal strain energy method;*
   - *Residual force vector method;*
   - *Modal updating methods;*
   - *Frequency response function;*
   - *Statistical methods.*

5. List important factors that influence maintenance of infrastructure, in general.

   - *Importance of the structure itself;*
   - *Maintenance cost;*
   - *New demand on the structure due to additional loads, if any.*

Tutorial Paper 4: Key

Answer all the questions (Total marks: 20)

PART A (5 × 1 = 5)

1. *Initial characterisation* provides the base line for comparing response of a structure before and after damage.
2. Significant changes in the vibration characteristics can lead to *damage localisation*.
3. Expression for damage sensitivity is not applicable for structures with *multiple damage locations*.
4. Both the crack location and change in size depend on the changes in *mass and stiffness* of the structure before and after the damage.
5. Significant changes in *eigenvalues* can enable a convenient damage detection process.

PART B (5 × 3 = 15)

1. What is the basic feature of vibration-based SHM?

   Changes in structural characteristics, such as mass, stiffness and damping caused by presence of damage, will affect the global vibration response of the structural system.

2. List the steps involved in vibration-based SHM.

   - *Measurement of structural dynamic response in terms of acceleration and displacement;*

- *Characterisation of initial structural model through both static and dynamic sets;*
- *Continuous monitoring and damage localisation of the structure;*
- *Detailed finite element analysis to update the structural model with the input from observed damages;*
- *Evaluation of the structural performance of the updated model.*

3. Draw the flow chart describing the vibration-based monitoring.

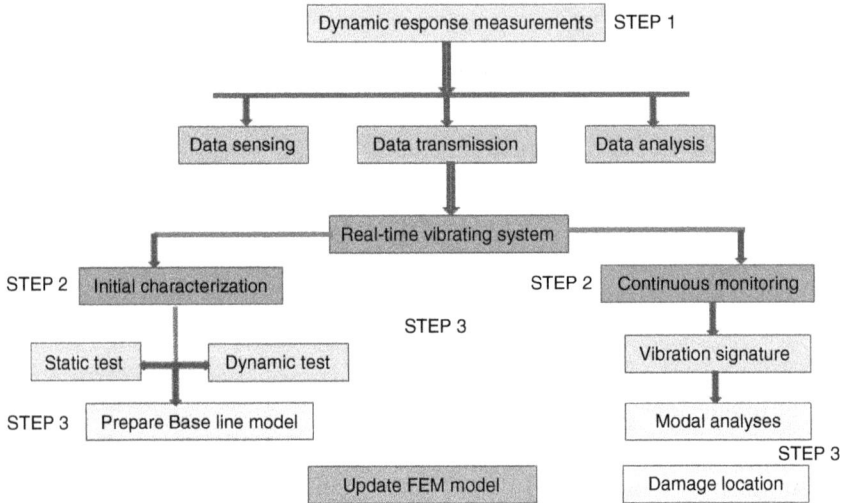

4. Define modal sensitivity.
   The ratio of the modal energy of the ith mode contributed by the jth member is called modal sensitivity.
5. What is the basic assumption in the method used to identify damage by element modal stiffness?
   Modal sensitivity for ith mode and jth member remains unchanged before and after damage.

Tutorial Paper 5: Key

Answer all the questions (Total marks: 20)

PART A (5 × 1 = 5)

1. Visual inspection methods affect *decision-making process and resource utilisation*, significantly.
2. *Anemometer* is the sensor used for wind load assessment.
3. *Natural frequency-based* methods are less contaminated by noise data.
4. Fractional change in the modal energy is related to *fractional change in frequency*, which occurs due to *damage*.
5. Curvature of mode shape can be approximated using *central difference* technique.

PART B (5 × 3 = 15)

1. The results of visual inspection (VI) may be inadequate to compare with true assessment. Why?

   - *VI teams may not be experienced.*
   - *VI guidelines used by different agencies may differ.*
   - *There are no standard guidelines for VI.*

2. List the factors that affect system complexity in SHM.

   - *Size and complexity of the structure;*
   - *Functional characteristics of the structural system;*
   - *Remaining service life of the structure.*

3. How can the maintenance of the SHM system be reduced?

- *Reduce the system redundancy of the structure.*
- *To avoid breakdown, provide renewable power source to the hardware of the SHM system. This eliminates the need to change the batteries in case of wireless sensors.*
- *There is a need to employ adequate IT professionals to ensure ongoing functional condition of the SHM system.*

4. What are the limitations of SCCM?

- *It is very difficult to compute the natural frequency with high accuracy.*
- *Hence, applying the correction based on its correlation to auxiliary mass is difficult and complex.*

5. What are the different stages involved in the long-term SHM?

- *Identification of structure;*
- *Risk analysis;*
- *Identify responses;*
- *Design of SHM and sensor layout;*
- *Installation and calibration;*
- *Data acquisition and management;*
- *Data assessment.*

Tutorial Paper 6: Key

Answer all the questions (Total marks: 20)

PART A (5 × 1 = 5)

1. *Love waves* are guided waves that travel in layered material.
2. PWAS is an active sensor. State True or False. *False.*
3. In flat plates, ultrasonic guided waves travel as *lamb waves* and *shear horizontal waves*.
4. In the pitch-catch method, flaws are detected by *wave dispersion* and *attenuation* due to damage.
5. Cracks are generally initiated in composites by *fabrication imperfection* and *inability to resist fatigue loads*.

PART B (5 × 3 = 15)

1. List the drawbacks (if any) of deploying ultrasonic method.

   - *Sound path traverses only on a small portion of the material volume. Hence, transducer must be moved to cover a large volume which is time-consuming.*
   - *Ultrasonic waves cannot be induced normal to the surface of the structure.*

2. Explain passive SHM.
   It uses passive sensors that are monitored over a period of time. The monitored data will be useful in updating the system characteristics.

3. List the advantages of PWAS?

- *Light in weight*;
- *Cheap*;
- *Simple and thin*;
- *No obstruction to the surface.*

4. Discuss the applications of embedded NDE using the pitch-catch method.

- *Corrosion detection in metallic structures*;
- *Damage detection in composite materials*;
- *Detection of debonding in adhesive joints*;
- *Detection of delamination in layered composites.*

5. Express the relationship used to determine probability of crack detection by the pitch-catch method.

$$P(\text{crack detection pitch catch})$$
$$= \frac{\sum \text{crack recorded by pitch-catch method}}{(M - N) + 1}$$

where M is the number of crack events recorded by NDE method and N is the number of serial events.

Tutorial Paper 7: Key

Answer all the questions (Total marks: 20)

PART A (5 × 1 = 5)

1. In the very early stage of monitoring, *embedded* sensors are used.
2. The period of monitoring during reconstruction is *4 times per day per sensor for 24–48 hours.*
3. Fibre optic sensors use *electromagnetic interference* to read and measure data.
4. Magnetostrictive sensors emit ultrasonic energy with very high strength. State True or False. *False.*
5. MEM sensors are manufactured using *very large-scale integration (VLSI)* technology.

PART B (5 × 3 = 15)

1. List the structures which require monitoring during the construction stage.

   - *Offshore structures;*
   - *Structures which are expected to face foundation settlement effects;*
   - *Near fault lines of seismic signals.*

2. How can the FOS be used in the measurement of moisture ingression?

- *It consists of swellable polymeric fibre optic sensors used to measure distributed moisture formulation.*
- *This sensor works in combination with optical time domain reflectometer to determine the spatial location of moisture ingression.*
- *This measures or identifies the point of moisture ingression by the attenuation principle.*

3. What is the magnetostrictive effect?

Ferromagnetic materials when placed in magnetic field are mechanically deformed. This property is called the magnetostrictive effect.

4. Mention the advantages of smart sensors.

- *They have the ability to continuously monitor the integrity of the structure in real time and can provide improved safety to public, particularly in case of ageing structure.*
- *They can detect damage at an early stage which can reduce the cost of repair and also the shutdown time of the structure.*
- *They are helpful in predicting the initiation of damage.*

5. Why is SHM mandatory for offshore structures?

- *Heavy mass construction;*
- *Huge capital investment;*
- *Downtime leads to loss of revenue;*
- *Prone to accidents, which is a threat to environment and human life.*

Tutorial Paper 8: Key

Answer all the questions (Total marks: 20)

PART A (5 × 1 = 5)

1. In areas where visual inspection is not possible, the members can be examined through *numerical model*.
2. The damage can be quantified before developing the sensing system by *numerical simulation*.
3. Wireless sensors convert the measured data into *digital form* for transmission.
4. In wired sensor networking, the central server is responsible for *collection*, *aggregation*, and *storage and processing* of data.
5. What is the most important advantage of wireless sensor network? *The amount of data that can be measured by the monitoring system is high.*

PART B (5 × 3 = 15)

1. List the factors to be considered while designing the monitoring system for offshore platforms.

   - *Sensors should be able to withstand environmental conditions.*
   - *The proposed SHM scheme should have financial advantage over the manual inspection method.*
   - *Vibration spectrum should remain stable over a period of time.*
   - *Normal sea state and wind excitation should be used to extract the natural frequency of the system.*

- *Above water measurements should be used to identify the mode shapes.*

2. What are the common factors that govern the selection of sensors?

- *Data format;*
- *Precision and accuracy;*
- *Linearity of data;*
- *Dynamic range of variables;*
- *Cross talk;*
- *Durability;*
- *Maintainability;*
- *Redundancy;*
- *Cost of the sensor.*

3. List the advantages of implementing SHM in strategic structures.

- *Ensures serviceability of the structure by long-term monitoring;*
- *Increases safety and knowledge about performance of the structure;*
- *Validates the design of the structure and its performance;*
- *Can monitor and control the construction process;*
- *Assess load capacity and therefore reduce the risk in the structure;*
- *Assess any requirement of emergency response effects.*

4. What are the limitations of wired sensors?

- *They cannot be implemented on large structures such as bridges, dams and offshore platforms.*
- *If the cable is damaged, data will be lost which leads to loss of efficiency of SHM.*
- *They do not have the capability to process data.*

5. List the advantages of MEMS sensor.

- *High accuracy;*
- *High reliability;*
- *Embedded in the structure;*
- *It has the capacity to measure the damage at early stage;*
- *Design alert monitoring system.*

Tutorial Paper 9: Key

Answer all the questions (Total marks: 20)

PART A (5 × 1 = 5)

1. One of the major challenges in the SHM design of offshore platforms is the *location* of the sensors.
2. BLSRP is *stiff* in the vertical plane and *flexible* in the horizontal plane.
3. What is the primary objective of the SHM design?
   Arriving at a layout of sensors costing less and functionally efficient.
4. *Microcontroller-based sensing units* are used to assess the errors in the collected data.
5. ANN combined with pre-processing tools, such as *Damage Relativity Analysis_Technique,* can be used for damage detection.

PART B (5 × 3 = 15)

1. Why is VI not possible in offshore structures?

   - *Inaccessible;*
   - *Hostile environment;*
   - *Characteristics of the platform is continuously changing.*

2. What are the components of the SHM system?

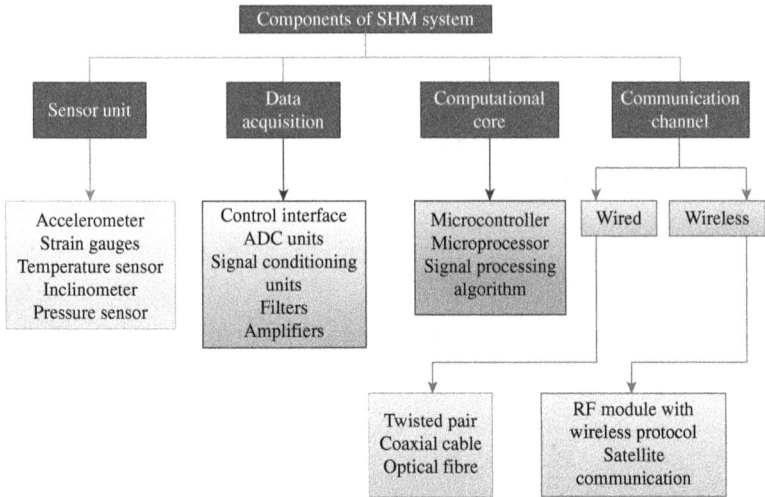

3. List some of the issues related to vibration-based damage detection.

- *Noise measurements and signal-to-noise ratio;*
- *Discrepancy between scaled down model and prototype;*
- *Nonlinearity in the structural response;*
- *Dense distribution of sensors;*
- *Influence of environmental factors in real time, which cannot be considered in the lab scale.*

4. List the design parameters to be considered in the wireless communication channels.

- *Data rate;*
- *Space range;*
- *Encoding reliability;*
- *Radio band.*

5. What are the advantages of GFRP?

- *High strength-to-weight ratio;*
- *Corrosion resistance;*
- *Military application like minimising electromagnetic radar signature on underwater vehicles.*

Tutorial Paper 10: Key

Answer all the questions (Total marks: 20)

PART A (5 × 1 = 5)

1. What is the main aim of embedded sensors?
   Decentralising the sensor fault detection and making it completely autonomous
2. *Data mining* can enable the pattern for future prediction.
3. Unsupervised learning is very appropriate to handle SHM of complex system. State True or False. *False.*
4. What is the primary objective in the conceptual design of the SHM system?
   To assess the vital requirements of health monitoring of offshore compliant structures.
5. *Microcontroller* is used to process the acquired signal from the accelerometer module.

PART B (5 × 3 = 15)

1. What are the salient features of machine learning?

   - *Robust control*;
   - *Human–computer interaction*;
   - *Speech recognition.*

2. What is data mining?

   - *It is the process of detecting patterns and structures within the data.*
   - *It is the broader concept of pattern recognition necessary in SHM.*
   - *It is a continuous and automatic system.*

3. List the factors that affect the output of the SHM system.

   - *Sensor calibrations;*
   - *Interpretation of threshold values;*
   - *Wrong pattern recognition.*

4. What are the issues that may arise in using wireless sensors over a large area?

   - *Interference;*
   - *Path loss;*
   - *Reflection.*

5. What are the sources of noise in signal?

   - *Presence of thermal noise;*
   - *Electromagnetic interference;*
   - *Sensor oscillation;*
   - *Quantisation of noise.*

Tutorial Paper 11: Key

Answer all the questions (Total marks: 20)

PART A (5 × 1 = 5)

1. In wireless sensor networking, power consumption can be reduced by decreasing the *number of samples.*
2. *Very Small Aperture Terminals (VSATs)* can enhance the communication efficiency because they can handle *higher bandwidth.*
3. *Alert Monitoring System (AMS)* generates the reliable report when the acquired data exceed the present threshold value.
4. The tethers in tension Leg Platforms are slack moored. State True or False. *False.*
5. *Shutdown* power mode is used to reduce the power consumption in the SHM processing unit.

PART B (5 × 3 = 15)

1. How is the sampling rate chosen in SHM?

   - *It is selected such that the device is able to deflect the changes associated with the maximum signal frequency.*
   - *The signal should be reconstructed if the sampling rate is twice the maximum frequency component.*

2. What are the different power modes in the processing unit of SHM?

- *Runmode in which all the functionality of the core processor are powered up.*
- *Standby mode, in which the processor can quickly rise up and interrupt, till such time the core is shut down.*
- *Shutdown mode to reduce the power consumption.*
- *Dormant mode in which the core is powered down.*

3. What are the alternate sources of power supply for wireless sensors in real time monitoring?

- *Solar charger;*
- *Alkaline rechargeable batteries with voltage regulator.*

4. What are the advantages of using SHM designed with IEEE 802.11 protocol at an operating frequency of 2.4 GHz?

- *There is no packet loss of data because the TCP layer will handle the issue in the data layer itself.*
- *It can also detect any failed packets and retransmit them automatically.*
- *There is no mismatch of the data because the packets will be sorted.*

5. Explain the WSN architecture with a neat sketch.

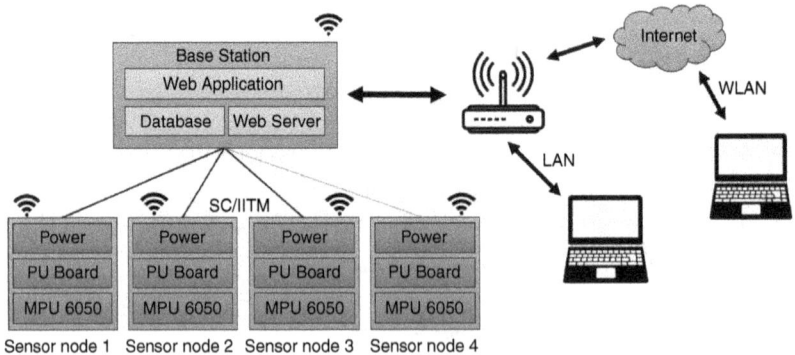

Tutorial Paper 12: Key

Answer all the questions (Total marks: 20)

PART A (5 × 1 = 5)

1. Frequency-domain technique is used to analyse *stationary event,* which is localised in *time domain.*
2. *Fast Fourier transform* is one of the best tools to identify the frequency component present in the signal.
3. What is the major defect in FFT?
   There is no localisation of the features across the time axis.
4. *Principal component analysis (PCA)* is the multivariate statistical model that can reduce the effect of environmental factors on the sensor performance.
5. Distributed health monitoring system is used in *massive structures.*

PART B (5 × 3 = 15)

1. How are postulated failure cases introduced in TLP experimental study?

   - *By applying eccentric load, which is a common scenario in offshore structures;*
   - *By removing tethers, implying the failure of tethers.*

2. Explain the feature extraction process.

   - *It refers to the processing of the time history data to extract sensitive damage features.*
   - *In case of dealing with large data from multiple sensors, these processors condense the data into small sets and then process them with statistical tools.*

3. What are the tools used to analyse the data in frequency domain?

   - *Fast Fourier transform;*
   - *Power spectral density;*
   - *Short time Fourier transform.*

4. What are the drawbacks of visual inspection?

   - *Time-consuming;*
   - *Expensive;*
   - *More complex;*
   - *Under water structures are inaccessible.*

5. What are the recent advancements in SHM technologies?

   - *A combination of natural excitation technique with Eigensystem realisation algorithm;*
   - *To carry out damage under dynamic excitation.*

Test Paper 1

Total marks: 100

Maximum time: 3 Hrs

PART A (20 × 1 = 20 marks)

Answer all the questions

1. SHM is a _____ involving capability to understand the importance of successful maintenance of _____.
2. SHM can be used as a new design tool in case of design of structures. State True or False.
3. _____ deals with the supervision of structures on a continuous basis using sensors in order to maintain the functional utility of the structure.
4. Structures with SHM deployment show a _____ reliability level indicating the assurance of acceptable quality of the structure in terms of _____.
5. In hybrid electromagnetic performing layer method, a sensitive magnetic field is created by _____, made up of _____.
6. _____ makes concrete more vulnerable to deterioration.
7. _____ is useful to detect the sub-surface nonlinearity caused due to material damage.
8. _____ is the non-destructive evaluation carried out to detect the defects in the structural system.
9. Local monitoring is helpful in determining the _____.

10. The scope of visual inspection methods is limited to damage detection. State True or False.
11. Damage with same severity occurring in symmetric locations will result in _____.
12. Lamb waves are vertically polarised. State True or False.
13. _____ is effective in detecting the fatigue crack propagation.
14. The magnetic induction of the material changes when the material is mechanically deformed. It is called_____ effect.
15. Sensors with _____ and _____ are smart sensors.
16. Through smart sensors, the sensing units are now capable of controlling the _____.
17. Delamination of GFRP leads to _____ and _____.
18. Smart sensors have _____ on board which makes them intelligent.
19. In WSN, dual microcontroller is responsible for _____ and _____.
20. What is the main aim of embedded sensors?

PART B (20 × 2 = 40 marks)

Answer all the questions

1. What is the importance of preventive maintenance?
2. What are the advantages of implementing SHM?
3. What are microwave sensors?
4. Differentiate damage and failure.
5. Give some examples of conductive composites.
6. Explain ultrasonic vibro-thermography.
7. What are the ways to handle uncertainties that arise in SHM?
8. Explain triggered monitoring with an example.
9. Differentiate local and global monitoring.
10. Define MDLAC.
11. What are the limitations of mode shape-based methods?
12. List the applications of PWAS.
13. Is monitoring necessary for a structure during special events? Why?

14. What are the functions of microprocessors in smart sensors?
15. What are the advantages of Wireless Sensor Networking?
16. What are the merits of sensor technology from the rapid technology advancement?
17. Name the sensors used to measure the kinematic quantities.
18. What are the factors which lead to major accidents in offshore platforms?
19. What is machine learning?
20. What is the aim of WSN architecture?

PART C (40 marks)

Answer all the questions

1(a). Explain in detail about the different types of sensors used in SHM. (7 marks)

 (b). What are the major concerns of using vibration-based detection in oil platforms? (3 marks)

  2. Describe the method to identify damage using element modal stiffness. (10 marks)

3(a). What are the limitations of frequency-based methods of damage detection? (7 marks)

 (b). Explain embedded phase arrays. (3 marks)

4(a). Explain in detail about fibre optic crack sensors. (7 marks)

 (b). Write short notes on ANN. (3 marks)

Test Paper 2

Total marks: 100

Maximum time: 3 hrs

PART A (20 × 1 = 20 marks)

Answer all the questions

1. SHM uses automated tools and system to improve _____ and _____.
2. _____ deals with controlling the dynamic response behaviour of structures under environmental loads.
3. _____ and _____ enhance the service life of the civil structural system.
4. Shearographic imaging is generated by _____, which is embedded on the surface of the structure.
5. _____ method is an alternate to a fully integrated electromagnetic technique.
6. Thermal gradients in the material are analysed to identify _____.
7. _____ monitoring is capable of identifying the major differences between the vibration-based measurements and environmental based changes.
8. _____ is defined as the validation of structural conditions.
9. _____ monitoring is carried out over the entire life of the structure.

10. Reference electrodes are used to measure inclination. State True or False.
11. In frequency-based methods, the damage detection strongly depends on _____.
12. _____ are useful in detecting the damage in thin plates and shells.
13. In the embedded method of NDE, transducers are temporarily inserted between the layers of composites. State True or False.
14. What is the period of measurement in continuous monitoring?
15. Based on the modification of light in the sensing segment, FOS can be classified as _____ and _____.
16. There is no transmission delay and data loss in wired sensor networks. State True or False.
17. BLSRP is _____ in the vertical plane and _____ in the horizontal plane.
18. One of the major challenges in SHM design of offshore platforms is the _____ of the sensors.
19. _____ can enable the pattern for future prediction.
20. _____ can be used to detect the existing damage.

PART B (20 × 2 = 40 marks)

Answer all the questions

1. What is the scientific justification of SHM?
2. List the components of SHM.
3. What are the advantages of SHM in the aviation industry?
4. What are the factors involved in operational evaluation?
5. List the common issues seen in concrete structures.
6. What is the influence of degradations in concrete as a construction material?
7. What do you understand by data normalisation?
8. What is the ambient vibration test?
9. What are the drawbacks in VI?
10. What are the factors affecting the inspection frequency of the VI methods?
11. What is the pattern recognition technique?

12. Explain the mechanism of PWAS.
13. Name the structures that are subjected to test loads before functional use.
14. Explain FOS as a single-point relative humidity sensor.
15. Why is it difficult to carryout VI or NDT in offshore structures?
16. What are the four functional modules in the SHM system?
17. What do you understand by computational core of the SHM system?
18. List some major accidents in offshore platforms.
19. What are the three categories of machine learning?
20. What are the reasons for multiple peaks in PSD plot of acceleration response measured in BLSRP?

PART C (40 marks)

Answer all the questions

1. What are the major challenges in SHM? (10 marks)
2(a). List the major advantages of SHM. (7 marks)
 (b). Write short notes on time reversal method. (3 marks)
3(a). List the specific objectives in vibration monitoring to estimate damage as per ISO (2002). (5 marks)
 (b). Compare conventional Artificial Intelligence and Computational intelligence. (5 marks)
4(a). How do you classify fibre optic sensors? (5 marks)
 (b). Write a short note on wired sensors. (5 marks)

Glossary of Terms

Accelerogram	Graphical out of Seismograph
ADC	Analog to digital convertor
AI	Artificial intelligence
ALARP	As low as reasonably practical
AMS	Alert monitoring system
ANN	Artificial Neural network
BLS	Buoyant leg structure
BLSRP	Buoyant leg storage and regasification platform
BPNN	Back-propagation neural network
CDF	Cumulative distribution function
CFGFRP	Carbon fibre Glass fibre reinforced polymer
CFRC	Carbon fibre-reinforced concrete
CFRP	Carbon fibre-reinforced polymer
CNT	Carbon Nano tubes
CPU	Central processing unit
DAQ	Data acquisition unit
DI	Damage index
DIC	Digital image correlation
DP	Damage prognosis
DRAT	Dame Relativity analysis technique
EMMC	Electro-Mechanical Mission computer
FBDD	Frequency-based damage detection
FBG	Fibre Bragg grating
FD	Fractal dimension
FDR	Flight data recorder

FFT	Fast Fourier transform
FOS	Fibre optic sensors
FT	Fourier transform
GFM	Generalized fractal method
GFRP	Glass fibre-reinforced plastic
GPIO	General Purpose input-output
GPS	Global positioning system
GPS	Global positioning system
HELP	Hybrid Electro-Magnetic Performance
HUMS	Health & Usage monitoring system
I^2	Invited inspection
I2C	Inter-integrated circuit
ICP	sensor Integrated circuit piezoelectric sensor
IT	Information technology
LNG	Liquefied Natural gas
LTM	Long-term monitoring
LVDT	Linear Variable differential transformer
MDLAC	Multiple damage location assurance criteria
MEMS	Micro-electro Mechanical system
MLP	Multi-layer perceptron
MPU	Micro-processing unit
MSC	Mode shape curvature
MSE	Modal strain energy
MSECR	Change in modal strain energy ratio
NDE	Non-destructive evaluation
NDI	Non-destructive inspection
NDT	Non-destructive testing (or) technique
OFDR	Optical domain reflectometry
P waves	Pressure waves
PDF	Power spectral density function
P_f	Probability of failure
PNN	Probabilistic neural network
PSD	Power spectral density
PWAS	Piezoelectric wafer sensors
RAO	Response Amplitude Operator

RCC	Reinforced cement concrete
RF	Radio frequency
S waves	Shear waves
SAMCO	Structural Assessment, monitoring and control
SAT	Site acceptance test
SCCM	Spectral central correction method
SDI	Single damage indicator
SHM	Structural Health Monitoring
SPI	Serial peripheral interface
SPR	Statistical pattern recognition
SSE	Safe shutdown earthquake
STFT	Short time Fourier transform
STM	Short-term monitoring
SWNT	Single-walled Nano tubes
TCP	Transmission control protocol
TLP	Tension Leg platform
VBM	Vibration-based monitoring
VI	Visual inspection
VSAT	Very small aperture terminal
WSN	Wireless sensor network
β	Target reliability index

Index